JeanJean

Photograph by Kate Hardie.

A Life of Unintended Consequences

Copyright © Malcolm Hart

All Rights Reserved

Published by H.H Dervish

A Life Of
Unintended Consequences

I never felt driven by ambition to be this or do that; I seemed content to let life find its own path. It's been a long path full of excitement, sadness, happiness, bewilderment, success and failure.

I've travelled afar; I've lived in different countries, had quite a few love affairs, marriages and other ill-advised forays into dangerous liaisons. I had career successes in advertising, fashion photography, film making.

A life of unintended consequences by Malcolm Hart

A Life Of Unintended Consequences

By

Malcolm Hart

Part 1: The way it ought to be?

1957. Eleven in the morning the doorbell is ringing. I'm on the divan reading. Not expecting anyone I don't move. Insistent knocking on the door. I give in. I roll off the divan, limp across the room and open it.

Two suited young men, one with a briefcase. They smile. The one with the briefcase politely asks if I am Meneer Hart. I nod. He announces they are from the Special Branch of Police. They show me credentials. I stand aside allowing them into my one-room studio apartment. They look around the room. I invite them to sit. I wasn't expecting them on this particular day but knew they would pay me a visit sooner or later.

The one with the briefcase does the talking. He's sympathetic; knew I was off work on account of a minor case of phlebitis in my left leg so took this opportunity to beard me in my den. Wishing me a speedy recovery, he opens his briefcase and takes out several sheets of paper.

Reading from them in accented voice more used to speaking Afrikaans than English, he enumerates every event in my life from the time I landed in Cape Town four years ago, a complete and altogether accurate account of where I've lived, where I've worked and the political radicals with whom I've associated.

He knew I had, during the four years, gone back to London returning to Johannesburg with a wife… who no longer lived with me.

The officer looks up; something else on his mind. There are one or two details he's uncertain about.

"Where is my estranged wife living now?"

Where is my estranged wife living? In contravention of one of Afrikanerdom's most sacred laws Jean is actually living with her lover Can Themba, the African deputy editor of Drum Magazine. Miscegenation, that means a Black and a White cohabiting, is regarded throughout white South Africa as a crime as heinous as treason or even murder.

I lie; tired voice sighs,

"God knows."

This was the first time she'd actually packed up and left me to move in with a lover. Until now she'd been satisfied with one-night stands.

Our marriage lasted the statuary seven years and then divorce. There was no resentment. She married Bill Oddie, fellow musician, broadcaster and ornithologist and they had two daughters Kate and Bonnie.

But it was not over for she and I. There were still warm, glowing embers and we continued to see each other. After her marriage to Bill failed, she married for the third time to Jim an intellectual Glaswegian teacher and philosopher. Jim drank himself to death. I think she was very much in love with him and his departure left her empty and lonely. She came through it but it left a scar. I guess it all added up, as I was to learn.

2.

I'm a South Londoner, born close by the Oval Cricket Ground, the third of four sons of a liberal-minded Jewish family. Dad is ambitious and hard working. He has a large hairdressing salon in Clapham and a small factory in Brixton that manufactures products for hairdressers; his premier product is a freestanding, domed, hair dryer he named Monarch; he also produces chemicals and paraphernalia for permanent waving, stuff like that.

Mum is warm, loving and honest; brooks no nonsense from the kids or anyone else; she divides her time between helping Dad with business and looking after us.

We live in a quiet, leafy part of a London suburb called Streatham; we have a house with a large garden and a tennis court at the back. We also have a bungalow on a beach estate on the south coast where we spend our school summer holidays. We have a car. By and large we're pretty well off.

For six years World War 2 filled the timelessness of my growing up. In the summer of 1939, as in previous years, we are on holiday at the bungalow. Young as I am, I detect a change in atmosphere around the place, the usual cheeky light hearted holiday banter has given way to an air of unease, the adults are discussing uncertain times ahead. The Nazis begin the aerial bombardment of Warsaw On the third of September of that year, while we're still down there, in accordance with our treaty with Poland, Britain declares war against Germany and the processes of mobilization began.

There's a strong surge of jingoism and patriotism, newspapers and wireless programmes tell us we are the best and not to be messed with; the

British Lion is shaking out his mane getting ready for the fray. In the village school where we're now temporarily lodged we sing from the heart, "There'll always be an England, and England shall be free…"

There are soldiers everywhere; gun emplacements are dug, barbed wire rolled out across the beaches. Gas masks issued and, breath held, we wait. I'm eight years old, excited by all that's going on around me, disappointed nothing's actually happening, no foreign soldiers invading our beaches to be mown down by our machineguns, nothing on the horizon, no enemy planes in the sky to test our anti aircraft preparations, guns snugly nestled in the emplacements I'd watched being dug.

Mum and Dad are faced with a dilemma. They fear for the safety of the children but their minds are also troubled by the safety of our home and business in London.

They hire a Nanny to look after us and we remain in the relative safety of the bungalow. Nanny is Scottish, takes her job very seriously and is excessively strict. We called her Battle Axe. Mum and dad thought she was exactly what was wanted; we would be safe with her.

They stayed on for a few days, then, with a shrug of the shoulders, more a salute to fate than anxiety, Mum and Dad return alone to London.

I guess they feared the worst, arriving back in London, but they find very little is happening. Air raid sirens are being tested and there are a few light air raids far away from Streatham. Everything seems to be near to normal. so they decide to bring us home and get us back to school and to as normal a life as possible. They have little idea of what's going on. Nobody does. That was the worst of it. The waiting. Not knowing what we were waiting for. We wait. We're in London nearly a whole year waiting. We go to school not knowing. Perhaps the war has fizzled out with barely a shot fired.

3.

It had not fizzled out. On the Continent the German army had reached Paris. In 1941 the Luftwaffe turned its *Blitzkrieg* on London, dropping thousands of bombs; not only on London, on all our great cities. The only good news was a government announcement that should an air raid last beyond midnight, school would be cancelled the following day. Fantastic.

In the relative calm of the days we children search blitzed streets for war's detritus, spent bullets, bits of aluminium blasted from fallen aircraft, valuable swap stuff.

It's the nights when the action really began; the throb and drone of heavily laden aircraft, punctuated by the whine of sirens, searchlights criss-crossing the night sky looking for them, anti aircraft guns blasting away at them as their bombs exploded seemingly haphazardly around us.

Now in real fear for our safety, Mum and Dad hurriedly evacuate us, older brother Basil, younger brother Robert and me. Oldest brother John stays on in London; he'll soon be of military age and drafted. The bungalow is not the best option for our evacuation, dangerously on the south coast where invasion might be expected.

Dad drives us to Tilsworth, a village in the Bedfordshire countryside about forty miles north of London. Going door to door, he and mum find families to board us. Young Robert, not yet five years old, is billeted with a warm family of chicken farmers in the village; Bas and I are left to share a bed at the home of a modest middle-aged couple and their teenage son. With their kids satisfactorily settled, heartfelt farewells and Mum and Dad are off, back to London.

Without the noise of sirens and exploding bombs, life returns to some kind of normality. At night I can see London burning on the reddening horizon.

We make friends with the local village boys. We divide into two groups and play chase across the fields. At night we'd call out to each other, "Holler, holler 'r' the hounds won't foller!" At weekends we'd open five-bar gates for sixpence to let the local gentry's hunt and their hounds pass through. I got a Saturday morning job muckin' out a local farmer's pigsty for three pence for a days work.

A few months later, Bas and I move to Dunstable four miles away where we're enrolled as borders at Dunstable School, a minor Public School. With all its teachers of military age away in the war, its academic standard is not what it might have been. Courage, sport and gentlemanly behaviour are important and are conscientiously encouraged along with some lesser kind of education.

Being a Jew at school can be a problem. Children can be unkind to each other and are quick to derogatorily label others and then fight them on account of their difference. But two-fisted brother Basil scares away any bullying anti-Semites. He has a formidable reputation. He isn't much good in the classroom but on the sports field he's a star and receives the entire school's affection and respect on account of it. And so it remains until he leaves to be swallowed up by the army.

The army suits him and within a month or two he's graduating from OCTU as a smart young subaltern. He looks fantastic in his new, well pressed, uniform. Sometimes he wanders Whitehall to have the Horse Guards salute him.

By this time I'm a privileged senior at school and less in need of his protection, I'm permitted on Sunday to go into the town to attend Evensong at the local Anglican Church. Perhaps the House Master thinks it might be beneficial, make a better man of me or even encourage me to become one of them. But Evensong for me is about flirty girls from Luton Girls Grammar School who would generally be there for no more religious purpose than my own. Cherish the thought.

Whatever I think are my motives, I'm entranced by the theatrical mysticism of the Church, the stained glass and incense, the peacefulness, voices echoing back off vaulted ceilings. I find myself spontaneously joining in exultant singing of timeless Christian anthems driven on by grand organ music. It touches something in me. I'm part of it. It feels good.

4.

Growing up at home as one of five males then attending boys boarding school for six years I learn little about women. Mum is the only female I know. I'm ten years old before I discover, through sliding down a rope in the gym, the connection between my penis and pleasure, I grow into adolescence knowing nothing about sex and ashamed of masturbating.

Dad asks if there's anything I'd like to talk to him about. He's too shy to raise the subject of sex himself. I know what he means, he's being fatherly but I'm not up to discussing it either. He and I, like most at that time, are still afflicted by a Victorian morality that induces shyness and a sense of shame around sexuality. Even serious discussion of sex is taboo; scatological jokes are OK amongst men. That, more or less, is where I have got to when Jean and I first meet.

5.

She comes from a different part of London; the docklands in the East End. Without siblings, she's the misbegotten daughter of a mostly absent merchant sailor and a flirtatious, self-interested woman; not even family enough to be called dysfunctional. The war and the bombing took her away from the attention they were failing to give her and, as an evacuee, enjoyed a better, friendlier passage amongst kindly strangers and schoolteachers. At the end of the war, about fourteen years old, she was returned to London to be reunited with the poverty and uncertainty of a mother she barely knew and an absent father. Word has it her father 'jumped ship' in Australia.

The part of London where her Mother now lived, around London's docks, had been heavily bombed and, having a daughter to look after, she became eligible for a *prefab* house supplied by the Government to those in dire need. It wasn't a good time for Jean, trying to adjust to the ways of her self-centred libidinous mother.

School was her salvation. She has a smart mind and school provided her with a means of escaping the fog of ignorance she lived with at home. Winning a scholarship to East Ham Grammar School for Girls, she was one of the first generation of working class East Enders to pick up a decent education and, almost inevitably, a confirmed belief in Marxism.

6.

We first meet in 1950 as students at St Martin's School of Art on the Charing Cross Road. I've already been there a year and she's part of the new intake. Nineteen years old, strikingly attractive, dressed in black skirt and sweater with a red neckerchief. She has everyone's eye, can favour whomsoever she pleases.

I'm fiercely attracted to her but don't know what to do about it, I don't know where this sense of urgency comes from. I desire to touch her, hold her close, kiss her. I don't know what to do? Where is this going? This strong sexual attraction just wells up in me, adding uncertainty to my confused mind. What's the next move? What if she doesn't feel the same way towards me? I'll never know. I daren't ask. I want her so badly, I want to take her to some foreign secret place and give in to my passion but haven't the courage to make any such move.

She smiles at me when we happen to pass each other in the corridor. I smile back. She's friendly enough whenever we find ourselves alone. There's plenty to talk about discovering more and more shared interests.

When she enters the common room and heads for me I'm chuffed. Her company is special; and sought by many. I'm relaxed with her. We share a passion for blues and folk music and hum tunes to each other. She has a guitar and teaches me a few chords. We sing together. I learn to play the guitar and harmonize with her amazingly strong, well-pitched voice. We become popular entertainers amongst our peers. We fall in love. I'm floating. This is forever.

7.

From its outset my relationship with Jean is dogged by the ease with which she's drawn to other smiling, admiring men. It seems she's attracted to anyone offering her love. I don't understand it. It's not the way it's meant to be. How could she declare her love for me yet at the same time bestow it on someone else? What's this love thing all about? Words spin around in my head but can't find sufficient coherence to express the depth of my disappointment. I'm too shy to discuss it with her. I'm suffering the first painful pangs of jealousy.

8.

My life with Jean took a turn. I'd become uncomfortable, depressed when I'm around her, fearful she might find someone more interesting than me. She seems not to notice. I've become suspicious of everything she says or does. At parties I watch her whenever she's in conversation with a man. I'm checking out her moves. Jealousy knows no bounds.

Monogamy has no meaning for her. I don't think she gives it a thought. She's being who she is. She knows it pains me and, wanting to change, continually protests her love for me, wanting all the things we fantasize about when we're in harmony, when it's forever. But there's something that stands in the way. Never having received it from her parents perhaps she doesn't know anything about love. Perhaps, in her growing up, free sexual behaviour was the norm. Is that the way it's supposed to be?

Maybe I'm old fashioned, although I don't think so. I'm of the mind there's something about partnerships that must be exclusive and respected. I'm expecting our relationship to mirror the only one I know, I'm expecting to settle into a deep warm and trusting friendship with Jean like the partnership my Mum and Dad have but it's not happening.

I love her beyond reason. Want her exclusively for myself. I don't want anyone else sleeping with her. Despite my despair over each of her short-lived affairs, real or imagined, I'm determined to marry her, sure that things will be different when we're husband and wife. Things will be as they ought to be.

I meet and admire her libidinous mother, a handsome, unhappy woman; spoken and unspoken frustrations constantly erupting into anger. She seems to accept me for who I am, her daughter's art school boyfriend, and there's no quarrel between us. Now and then she'll cook us sumptuous roast meals and we spend occasional evenings with her in the pub close by the George V Docks where she sings in strident cockney voice popular songs of the day to an admiring crowd of dozy drinkers.

The only time I meet her father is when he's ashore visiting her and fucking her flatmate. He only spoke to me once. Looking me hard in the eye with his sea-faded blue unblinking eyes he says,

"Jews are a canker on the human race."

9.

Mum and Dad, being modern people, subscribe to a modernised version of Judaism, Liberal Judaism. Their rented synagogue used to be a Baptist chapel; a hall with a platform at one end. Much of Orthodox Hebrew protocol has been abandoned; men and women sit bareheaded together in chapel-like rows; the Rabbi preaches from the platform below the *Sefe Torah*, the Sacred Scrolls, housed in a cupboard fixed to the wall behind him and conducts the service from there in English.

I'm unenthusiastic about attending. It's a boring duty. It doesn't resonate with me. We only attend service on High Holidays, weddings and Bar Mitzvahs. I go along to please my parents. There isn't the palpable mysticism I'd enjoyed in the Anglican Church at school. No beautiful stained glass depictions of biblical events; no incense; none of the singing to robust organ music. I love the Christian Church's atmosphere but can't in any way accept its vital tenets. I don't really embrace Judaism either. I don't find organised religion inspirational.

Mum and Dad may be modern folk but memory of their parent's cultural habits linger. There's still a consciousness of *Kosher*. Eating *Kosher* for us is a weekend practice, a trip in the car to the East End, a kosher chicken from Kaplan's, a *challah* from Kossov's and, a cream doughnut for me if dad thinks I'm deserving. It had little to do with religion. It was more the pleasing shadow of a long-held cultural habit.

10.

One day in 1950, young brother Robert returns from school with a cyclostyled letter from the Head Master; he hands it to Mum. It's a list of names of young men from a high school in Johannesburg, South Africa. They'll be visiting London this summer and are looking for host families to take care of them for a couple of weeks.

South Africa? Johannesburg? Distant place names that mean nothing to her. She doesn't know anyone there; no family there, not even friends. There's no reason for her to have ever given the place a thought. She runs her eyes idly down the list. They hover over the name Raymond Rothschild.

There's still a whiff of anti-Semitism in the English air that generates and sustains in Jews an unspoken feeling of care for each other. *Landsmannschaft*. Mum's eyes hover over the name Rothschild and settle there. Raymond Rothschild, a *landsman*.

Raymond arrives that summer, a short, dark-haired, bespectacled seventeen year-old. He's well mannered and amiable. He goes off every morning to join his classmates on tours of London so I see very little of him. A week later his young sister, Beth, arrives with their older brother Alan. Beth is fifteen, as amiable, dark-haired and short as Raymond. I don't see much or them, either. There isn't the time. It's summer 1950 and the spirit of the Festival Of Britain has taken over the South bank of the Thames. Jean and I are occupied after school every evening working in a floating restaurant moored outside the Royal Festival Hall. While she's serving on deck, I'm washing dishes down below.

Then Raymond's parents arrive in London. Raymond leaves us to join the rest of the family at the Rembrandt Hotel on Piccadilly. Bernie Rothschild and his wife are a respectable, undistinguished-looking elderly couple, perhaps ten years older than Mum and Dad. Moved by the warmth and kindness their son has received in our home, they want to show their appreciation. They invite Mum and Dad to spend time with them in their home in Lower Houghton, an exclusive, leafy suburb of Johannesburg.

11.

Travelling by sea, Mum and Dad are away for three months. They return, tanned and in great spirits; they tell us Johannesburg is the most wonderful place on earth; friendly people; splendid climate; no post war rationing and plenty of career opportunities. They want us all to move there. The racial intolerance that infests the place already well known to me isn't included in their glowing narrative. I know I couldn't live in a place like that. The entire family is so excited, full of the spirit of the adventure; not the time to tell them I'll not be going with them.

Nor could I tell them I'm in love with a *Shiksa*, my non-Jewish Jean who I'm determined to marry. Even progressive Jews of their generation retain firmly negative beliefs about mixed marriage. Whatever their feelings, I'm not going to South Africa or anywhere else without her. I hide my refusal behind the importance of continuity in my life and work already started at St Martin's.

They're disappointed, as I knew they would be. I, too, am sad at their imminent departure but at the same time excited at the thought of freedom. I don't want to live in a country that has a colour bar. I lie to them. I'll join them when I graduate. Later that year Mum, Dad, my three brothers and sister-in-law with her new born daughter Stephanie, ship out.

Before leaving, Mum and Dad, having sold the house and its contents, want to find me a suitable place to live. My cousin Peter, five years older, veteran Fleet Air Arm gunner, has just graduated from law school. We've always been good friends beyond our family ties. He, too, is breaking away from home and traditional Jewish family values. Even so, he's considered by Mum and Dad to be a stable, responsible adult; he's a lawyer, after all and they think it a good idea for us to share a flat.

They arrange for us to meet one of their friends, a painter, a devout Christian spinster who rents rooms to students in her large Clapham Common Old Town house. Peter and I visit her. She welcomes us and offers us rooms but the social restrictions she would impose are unacceptable and we politely decline.

Dad's angry.

"Do we want to live in a brothel?"

Peter replies,

"No. Nor do we wish to live in a convent."

We find a second floor flat in a large house across from Streatham

Common. I introduce Peter to Jean and all my closest art school friends, most of them older than me, some having got caught up at the end of the war like Pete. He takes to them and they take to him. They like him for his easy hospitality and, when asked for, his freely dispensed legal advice. It's a haven for Jean and me. She doesn't move in but we spend much of our time there.

12.

1952. A three-month summer break from Art School. Mum and Dad have invited me to come to Johannesburg to celebrate my twenty-first birthday and hopefully, I guess, change my mind about living there. I set sail from Southampton for Cape Town on the Winchester Castle, a sun-splashed, lazy, two-week adventure. Hot days become warm evenings of drinking, flirting, dancing to a jazzy trio after a dinner of considerably higher quality than the rationed fare at home.

I become friendly with Pamela, an English girl on her way to Cape Town to marry. On the last night as we enter Cape Town's harbour we stay up to watch dawn slowly revealing the weighty mass of Table Mountain and the myriad lights of houses all the way down its lower slopes.

In Cape Town I board the overnight train to Johannesburg. I lie in my bunk listening to the beat of the train across a thousand miles of mountains and deserts, dozing off watching the lights of small towns appearing then disappearing back into the night on this seemingly endless journey.

It's the middle of the Southern Hemisphere's winter. It's cold. Johannesburg sits at an altitude of six thousand feet; the sun, brilliant as it is in a cloudless blue sky, fails to warm the air. Basil meets me at the station. We get into a large American car, a Plymouth saloon belonging to Dad; we drive through the busy streets of the city and out into the suburbs to the new family home. We park outside a recently built, modern concrete block of apartments, its hard lines softened by surrounding areas of bedded plants, grass and trees, distant from the noise of city traffic. Not a bad place.

We haven't been together as a family for over a year and our reunion is a celebration. We're happy to be with each other again and stories abound. What's been happening? They're eager to tell me. Dad has opened a small chain of successful ladies hairdressing salons. John and Basil work with him, Robert is still at school. They talk about me joining them. Why don't I come? Dad has investigated the Art School here at the Witwatersrand University. It's said to be very good. They whet my appetite with stories of game parks and fabulous beaches they've visited. They're all happy with the day-to-day life, the weather, everything. No complaints. Can't wait for me to join them. I nod. They say nothing about apartheid.

13.
They've become more Jewish; I don't understand it at first but recognise it as the sub-conscious pressure of living in a raw racist society; sub-cultures fearing for their own security, drawing in closer to their own. The family attends the Reformed Synagogue on high and holy days as they did in London. It's a little more formal than the Liberal Synagogue regarding orthodoxy. In the new reformed synagogue, heads must be covered. Amongst the worshippers, I detect a more tribal feeling than we

experienced in London.

This new Jewishness manifests in the home. Dad has fixed a *Mezuzah* to the frame of the front door. We never did that in England. It wasn't smart to publicly identify one's self there as a Jew. In Johannesburg, it's the custom. All their friends are Jewish. It's a little Israel. By and large, they seem to express a stronger affinity with Israel than with any of the governing bodies of their own country. With holocaust still fresh in the mind, unconsciously they seem to be investing in Israel their hope for a safe future.

I spend time with the young Rothschilds and their friends. Apartheid? They don't give it a thought. They were born into a segregated state. It's their norm. They protest they treat their natives well.

Their social style reflects the American movies they see at the *Bio* when they're best dressed for a date on Saturday night. They live in a teenage world in which the pursuit of sex dominates. They call each other by affectionate slang names like 'Cup Cake', 'Babe' and 'Hun'. They've never met an African man to man, only as a servant. Politics doesn't interest them, doesn't enter their conversation or their lives. Don't let's spoil the fun.

Segregation is an institution taken for granted. In all public places there are separate entrances, clearly marked, one for Whites only and the other for anyone of colour collectively called non-Europeans. The same notices appear on elevators, one for Whites only and one for goods and non-Europeans. In the park there are leisure benches designated in the same way. Nobody comments. Whites don't notice any Black resentment. It's the way it is.

We spend polite time as a family with the Rothschild's family. I haven't seen the elders since their visit to London two years ago.

Raymond has become engaged to a very attractive, nineteen year old blonde. Beth, now about seventeen, has made the catch of the day, Abe Brody, the currently most desirable, most handsome, recently graduated doctor from a family wealthy as her own. They will soon be married.

14.

I'm bemused by these relationships. Outwardly, there seem to be no shared interests between them other than other upcoming weddings, where to buy houses, furniture, carpets and curtains. There's no display of fondness that I expect from people who have decided to marry. The attraction is obviously beyond physical. There are other considerations besides passion that determine the suitability of a marriage partner. Parents as well as their children have dowries and inheritances on the mind. It's not a shameful secret. They talk about it openly checking out the odds. I guess, being a young Socialist, this consideration of wealth in a marriage disturbs me but it's beyond criticism. It's tribal. The most traditional native African families take account of the wealth of those their children are marrying.

Sunday morning I'm taken to visit a gold mine. Every Sunday certain African miners get the day off to give performances of their tribal dances to an audience of tourists. Each mine employs Africans from different parts of the Union. There are Zulu and Xhosa, Hausa and Himba. Their dances vary accordingly. Dressed in animal skins and other beaded tribal accoutrements they each perform their tribe's traditional dances. As a finale they throw in a modern innovation, the Gumboot Dance, a Miner's tribal dance.

Changed back into their mining clothes, helmets and the all-important gumboots, the dance involves much leg raising and stamping interspersed with rhythmic hand slapping on the boots. It's angrier than the traditional dances. I see the release it gives them, wildly expressing themselves in the winter sunlight in a way only they know how, these lusty men who spend their lives shovelling and drilling thousands of feet underground for little gain.

15.

A school friend of Raymond Rothschild, Chaim Snipelespki, invites me, to spend a few days on his family's farm in the Eastern Transvaal outside the small town of Bethel.

A luxurious farmhouse, spacious, with all the comforts you might expect to find in a five-star hotel. I'm treated like an honoured guest. Food and wine are abundant, the atmosphere warm and friendly. I see little of my host's father. Running his large farm doesn't leave much time for him to socialise and he apologises for it.

On the farm itself, working in the fields close by the house is an entire labour force of African convicts hired out from the local prison, few of them serious criminals, most of them doing time on account of some minor infringement of the Pass Law. White prison guards mounted on horses, rifles cradled in arms, oversee them. Orders are yelled in coarse Afrikaans. When they break for food and water the workers feed themselves in the farmyard, filling their tin cups from a large trough of maize sludge.

16.

I'm back in London with no clear idea of where I've been or what I've experienced. Jean wants to know everything and I don't know where to begin. She's expecting some kind of judgement, some kind of political assessment of the situation in the dark, racist country.. I'm finding it difficult to analyse my feelings. It had been wonderful to be with the family again but, apart from the insult of apartheid, I'd felt uncomfortable with the polarisation of Whites into exclusive tribes. Afrikaners, Jews, the English-speaking anglophiles, all of them uncomfortable with each other, all of them supporters of the status quo.

My most enjoyable experience in the two months I spent there was the music, African music. I've brought back with me a number of vinyls, Gallo recordings of Solomon Linda and his Evening Birds, one of them the original Zulu version of the well-known song *The Lion Sleeps Tonight* called *Mbube*; and *Lizzie*, a popular romantic song by Alpheus Nkosi. Jean loves them and is quick to learn the words.

The only change while I was away is Jean's decision to quit St Martins and apply to a number of universities where she can pursue a more academic study of Art rather than practice the real thing. I wasn't impressed. I thought her a good draughtsman; her work was strong. And I'd miss her daily presence here at school. It did little for my fear of losing her to someone else.

17.

After four years at St Martins studying as an illustrator I'm unsure I will be able to make a living from a diminishing publishing need for that kind of work, work already well covered by talented artists like Ronald

Searle, John Minton, Bill Papas. The wisdom of the day is to get a teaching diploma just in case. There's always a need for teachers. I'm accepted on the pedagogue course at Brighton College of Art.

I share a flat overlooking the sea in Brunswick Square with two other guys on the course. I manage to get the single bedroom. Brighton's really quite a splendid place to be as is our apartment. The beach is across the road and, in the winter months the spray from stormy seas beats across our windows.

Another student on the course is the beautiful and talented Molly Thomas. Like all the other men on the course I'm curious about her.

Geographical as well as emotional distance has grown between Jean and I. she's been accepted by Reading University and lives above a boathouse, a beautiful flat overlooking the Thames at Sonning. She has new friends there and I'm far way in Brighton. When I visit her I'm jealous at how close she's allowed some of her new friends to become. She's lost to me in her new environment.

I fall in love with Molly. It's easy. She's a beautiful, talented painter and writer and so different from Jean, so much more honest. When we talk she holds my eyes with a frankness I can respond to. She means what she says. It's fresh. By now I'm more experienced in love's song and dance and my attentions are welcomed and reciprocated but before she commits to a deeper relationship she would like to see my work. I arrange to show her my portfolio like I'm applying for a job.

18.

Molly lives locally. Passed muster, I'm invited home, the whole family lives in a flat over her Mum and Dad's sweetshop on the London

Road at the edge of Brighton. Some evenings Molly and I take charge of the shop so her parents can relax for a couple of hours in the pub. They enjoy a drink. It's important in their lives.

I meet Moll's older sister Sally with her handsome rugby-playing husband, Granville. They live in Bridge End, South Wales, a friendly and proudly Welsh couple. We understand each other and, whenever they are visiting, take pleasure in each other's company.

After a few weeks, my relationship with Molly is growing less extraordinary and becoming more the norm. Certain personal incompatibilities are arising between us that neither of us seems able to overcome. We share love but it's not enough to resolve the situation. When we are together I'm experiencing a subtle change in the vibration between us a loss of confidence. Molly is saddened by it. She tries to fend it off; a ring would comfort her to show the world our commitment to each other. At a store in the Lanes I buy her a slender, silver thing set with a tiny red garnet and amongst our friends we celebrate the role of newly engaged.

19.

Her father refuses to accept the notion of our engagement let alone our marriage. He reveals himself a fascist and makes no bones about it. He deeply resents his daughter hanging out with a Jew and talking about marriage what the hell. His wife, who I believed liked seeing me and Molly together is as uncertain about his feelings as she is about most things; Molly, and Sally and Granville when they are there, throw the Dad's prejudices back at him, scolding him for his Nazi shit; he shouts back his crude racist rhetoric, thumps the table and slams out of the room.

The state of Molly's family life doesn't encourage my involvement in it. Molly's mother, a sensitive soul, can't manage it without a drink. I can't manage it. It started well but is foundering; the flower is dying on the stem. Christmas is upon us but none of the promised joy. In my heart I know this relationship is over; a feeling of emptiness and sadness. We really believed we were meant for each other but it wasn't to be.

20.

I've returned to London, staying with Peter for Christmas. He's arranged a party inviting all our friends. It's like old times. We haven't seen each other for a year. There's a lot to talk about, a lot of laughter, a lot of wine, a lot of love. Not surprisingly, Jean is there, too. I had half expected it, silently hoped for it. She seems as lost as I am. We chat; both acknowledge that something had gone missing in our lives and here we are together, loving friends again. We sing to our friends. Everything is as it should be.

21.

In England, conscription at the age of eighteen into one or other of the armed forces is the law. For those pursuing further education call-up could be postponed until after graduation. I'm graduating. Postponement is over. I have to check out my options. Growing up in the war gave me an appetite as a child for manly military pursuits, continued and reinforced later at Dunstable School which made the Army as a final career destination very attractive.

But five libertarian years at art school had certainly dulled the attraction. The idea of two years in a peacetime army had become as unattractive to me as going to South Africa, but like it or not, South Africa is now an option.

The conscription law carries a statute of limitations. I'd be free of the law after I reach the age of twenty-seven in four years time. The mind weighs the pros and cons of the situation. I discuss it at length with Jean, argue with myself about morality and societal responsibility but am beginning to give serious consideration to South Africa. I try to reassure myself it wouldn't be too bad. Might be interesting. Four years.

Jean's thoughts about going to South Africa fluctuate daily between complete Left-wing resistance to the idea and her feelings of love for me when she says she'll do whatever I decide. I tell her I'll go if she will come with me. She agrees to come in a year's time when she has graduated. It's not what I want. She proposes I return in a year, we'll marry and then go back to Johannesburg together. I'm unsure our relationship is strong enough to survive that long. I board the Union Castle liner alone bound again for South Africa.

22.

Being back with my family in Johannesburg, living with them in their large Parktown apartment, is like being back in a different world, a world I once inhabited, familial love I can depend on and enjoy. Separation for us was no big deal. During the war we'd been separated for the best part of four years, kids in a Bedfordshire school, parents taking care of business in London.

Robert is still at school. Bas, not short of choice, is waiting for the love of his life to appear, John, Sheila and their baby Stephanie have an apartment nearby. All are thriving.

I chat with Selma, Mum's help. Originally from Bechuanaland she's lived and worked in Johannesburg most of her life as a servant. Like all apartment buildings there's rudimentary accommodation for servants in concrete rooms on the rooftop and that's where Selma lives. She's amiable and used to working in White families. She's self confident, unperturbed by racial differences. For her, race laws belong to another world somewhere out there.

23.

I find my first job in an advertising agency, Van Zijl & Van Zijl, an Afrikaans company managed in large part by expatriate Englishmen. I show my art school portfolio to the Art Director, an English eccentric who, after serving in the war, had fled the ensuing uncertainties in Britain for warmer climes and fewer challenges. He likes my work and hires me questioning sardonically why I should want such a miserable job in such a miserable country.

The middle class Afrikaner people I work with communicate in a language I don't understand; it's the first time I feel I'm in a foreign country. Working among them is not comfortable for me, redolent of the feeling of exclusion I'd sometimes experienced in school as a Jew. They regard me as an outsider, not because I'm a Jew but because I'm an Englishman. They still feel anger at ancestral memories of the Boer Wars, their defeat at the hands of England's army, the dissolution of their society, concentration camps, the great trek north across the Vaal River to be rid of British rule. They're polite enough but if I overstep the mark, become too friendly and familiar with one of theirs, they politely elbow me away.

Now I have a job I can afford to look for somewhere of my own to live. I've found a first floor studio flat in a new block in Yeovil, close to the centre of town. It's a small space; living room, kitchen, bathroom, but the floor to ceiling, wall to wall window in the living room opening onto a balcony overlooking the street gives it a sense of space beyond its dimensions. At weekends I sit on the balcony sipping beer in the sun watching bare-footed African domestic workers at street level in their dark blue uniform shirts and shorts, taking time off chatting and smoking.

A man walks down the street playing a guitar to the rhythm of his walking; another strolls by playing an *mbira* (a thumb harp); sometimes a penny whistler strolls jauntily down the street. Music is important to them in their second-class lives.

One night I'm woken by a scraping sound coming from the transom window above my bed. A long pole, fed through the transom, is wavering over me. I look up and see the dark silhouette of a man outside the window wielding it. I scream. The pole and the silhouette quickly disappear. It's a method of burglary new to me; hooking clothes and whatever else while their owner sleeps. In the morning I contact a company to have the entire window made safe.

Returning from work I'm shocked by the installation of iron bars protecting the window. I can't believe the change it has made to my light, airy room. Now it feels more like a cage but I see the point. The ground and first floor levels of practically every domestic building in and around the city have ironwork on their windows. Nobody is fazed by it. It's become an architectural feature of the time. No one considers himself or herself to be under siege, just safer.

24.

I spend most Friday nights with Mum and Dad and friends of the family. Sometimes friends of Dad from the synagogue or business associates would be there. Although he enjoys living in South Africa, Dad has never bought into the generally perceived validity of apartheid. The colour of a man is not important to him and despite shaking heads and warning, wagging fingers from his White friends, he trusts and shows respect to anyone who works for him. He silently endorses my own views on racism and enjoys seeing me argue the toss with his friends.

One of them is telling me that as someone new to his country, I need to understand our naïve and rather slow-witted natives. He patronises them. They need us. We have to take care of them.

It doesn't look that way to me. I'm curious. I ask him if he, personally, knows any Africans.

Of course he does.

How well does he know them?

He's known John his gardener for nearly twenty years; he remembers, chuckling, how he had to teach him to climb a ladder.

What is John's family name?

He smiles. The name he called himself was so unpronounceable I changed it to John. He likes it. He always smiles when I call him John.

Is he married?

Probably. They sometimes have many wives don't you know. He winks; maybe that's why he smiles so much. Does he have any children? He chuckles; old John? I'm sure he does. They drop picanins faster than you can count. You can't blame them. They are primitive, simple people. It's just the way they are.

He wasn't exactly lying. He just didn't know any of the African people he lived amongst. Any Africans employed in his household as cook, cleaner or nanny to his children knows much more about him than he evidently knows about them. Most White South Africans share this ignorance. The *status quo*, this continuing disinterest in African's hopes and aspirations, suits them fine without a thought for the future. Their blindness to reality leaves them blissfully ignorant of an African intelligence and political awareness brewing around them before their unseeing eyes.

25.

My advertising work brings me into contact with commercial photographers who interpret the sketches I make for advertisements. I come to know Sam and Aleida Haskins. Aleida claims her family landed here with Jan Van Riebeck three hundred years ago. Despite their Afrikaans ancestry and general support for the government's line on apartheid, Sam is the best photographer in town and we become friends. Apartheid doesn't come into it.

Despite Sam and Aleida's reactionary views, I find them hosts to a small circle of young creative free thinkers in Johannesburg, none of them racist. Attracted by Sam's creative energy they find their way, sooner or later, into his studio.

Sam's favourite male model handsome Arthur Goldreich, artist, architect, arch political dissident, veteran of the Six Day War in Israel.

Aaron Witkin, gentle giant, a painter of considerable talent, in love with Sam's assistant, Colleen, is constantly there.

Blond haired, open faced Eduan Naude, a talented, gay, fashion designer as Afrikaans as Sam and Aleida but with none of their racial prejudice.

Spending time with them in their creative freedom makes me feel my life is boring, meaningless. What does my creative life amount to? At work I resize photographs to fit a satisfactory disposition on a page with text. It's work, limited in its scope and audacity by its commercial purpose. Through my friendship with these free, like-minded individuals, I maintain some kind of connection with Art. Besides my advertising work I get one or two commissions to paint murals in trendy cafés.

Occasionally an attractive nineteen-year-old Swedish girl would turn up at Sam's, Frieda Blumenberg, blond hair, blue of eye and clever at making jewellery. There's something mysterious about her, a shade of unhappiness, perhaps. She definitely has my attention. She's not given to publicly showing her feelings but seems to respond. We make a date. She suggests we go rock climbing in the Magaliesberg Mountains, about two or three hours drive north.

Slim and lithe, she springs from foothold to handhold like a mountain antelope, leaving me hesitantly crawling behind. She waits patiently for me to catch up then bounds on while I take a rest, smoke a fag and wait for her to return.

She's exhausted and leans her body against me in an affectionate and trusting way as we drive back to Johannesburg. I'm entranced by her boyish female beauty, her short tussled blond hair, her youthful skin, her sad blue eyes. Making love to her is adoration. I don't give Jean a thought.

26.

Saturday morning browsing in a record store on Kotze Street I run into John Goldblatt. We don't actually know each other but there's recognition. He's a big, full bearded man, a black beret covering a balding head; his serious face rarely wears a smile. I'd seen him several times in the Common Room at St Martin's with one of our students. We're both in South Africa for the same reason, both far from home dealing with life in a racist society rather than serving in the army. We become friends. His attitude to differences of race is not political; it's more existential like my dad, dealing with people on their merits without giving thought to the colour of their skin. He teaches English in a cram college in the centre of Johannesburg. John is cultured in a very rough and tough way, well educated and well informed. We spend a lot of time together; I can learn from him; he broadens my appreciation of literature and of classical music.

He's already made an African friend. He introduces me to Bloke Modisane. They met in the same record store, a rare meeting, not the kind of thing that happens everyday; a White man openly socialising with a Black person of either gender.

Bloke's a confident, educated man in his late thirties. He deals with local day-to-day racism with a degree of humour regarding it as silly behaviour. The Whites call any African 'John' without regard for the man's dignity. When anyone asks Bloke for his name he says John.

He has style, dresses like American newsmen you see in the movies, fedora tipped back on his head and walks with something of a John Wayne swagger. He used to work at Vanguard, Fanny Kleneman's left wing bookshop; by the time we meet he's the Social Editor of Drum Magazine.

In the atmosphere of apartheid, any kind of social intercourse with people of colour is under suspicion. Talking with an African in the street in broad daylight gains the concerned, critical attention of passers by and is likely commented upon or even reported to a policeman. I become cautious about whom I'm seen with, paranoid you might say but that's the way the crazy mind works when confronted by these ridiculous circumstances. It questions the trust between Black and White. Am I faking it? What does he want? Do they think I'm playing some kind of White liberal game? Am I? This smiling African friend is not smiling at all; he's waiting to cut your White throat. They joke about it. Comes the Revolution I wouldn't cut your throat *baas*. That guy over there; he'll cut your throat. It's ridiculous but on everyone's mind.

I'm beginning to feel just by being in South Africa my White presence is contributing to apartheid. I want to make it clear. I'd like to do something useful in the Black community; anything. Bloke takes me to the Bantu Men's Social Centre located at the sleazy edge of Johannesburg's central business district, among car dealerships and cheap food stores. It has a library, gymnasium and theatre stage and is home to the ANC's Youth League. A number of people there are struggling to learn guitar chords to give backing and rhythm to their songs. I go one evening each week and teach them what I know.

27..

Jean graduates. I return to Britain. The train to Cape Town is ready to depart, a small crowd of family and friends gather to see me off.

Frieda appears behind them. She doesn't know them, doesn't approach but stands off smiling wistfully at me. She knew this day would come. I'd told her I planned to marry. I'd like to be with her, I'm sad to be leaving her but foremost in my mind is the promised marriage to Jean. I feel no guilt but it's quite heartrending really. Freda's such a pleasure to look at and be with.

Back in London, Jean's pleased to see me, clings endearingly to me, but is still unsure about South Africa. She'd always been equivocal about spending the next four years there. She still bridles at the idea of living in a country that flouts people's human rights. She's a Socialist and the degradation of anyone on account of race is anathema to her. She really doesn't want to live there. When push comes to shove, after a year apart she's now equally dubious about committing to marriage.

It's a bitterly cold morning in January 1954, the day of the wedding. We're staying with Peter. Jean wakes up in a deep funk saying she can't go through with it. I'm angry as hell thinking of all our promises, the arrangements we've made, the people we'd invited. I threaten her with worse than death if she walks away. Later that morning we are married at Wandsworth Town Hall. The wedding photograph, a grey picture, looks more like a public execution than a wedding.

Jean is in a state of depression. Ten minutes into the reception, frustrated and miserable, she throws her wedding ring at me and flees. Everyone's astonished. I don't know what to think, all kinds of society's shibboleths invade my shocked mind. My ego is shattered. I want to hide.

It takes the best part of two days to find her and after a bitter row and tearful reconciliation she finally agrees to come to Johannesburg with me. Perhaps it was love or perhaps she was finally overcome by her own sense of adventure. Perhaps she had managed to convince herself South Africa was a politically interesting beast and should be investigated.

Before we leave I write to Mum and Dad telling them Jean and I are married. I receive a terse reply from Dad. You've broken your mother's heart and I cannot find it in myself to forgive you. The response is stronger than I ever imagined. Even so, I don't take it too seriously. It's like a diktat, not from his heart but something he thought was the right thing to say under the circumstances; a required traditional reaction. I never doubted his love.

28.

Jean and I are leaving for Johannesburg bound for Cape Town aboard the Caernarvon Castle. Passenger liners traditionally celebrate crossing the equator with a gay, light-hearted ceremony called "Crossing the line". That's what we are doing. We're crossing a line, we have no idea where crossing it might lead us.

Mum and Dad receive us affectionately. They're disappointed Jean isn't Jewish but will love her anyway and live with it. It had been difficult for their generation to accept mixed marriage. They were the children of the first immigration of Jews from elsewhere. Some still retain a sense of separation from the host community. They still worry. They'd seen early failures in the mixing of cultures. I understand but think concern over mixed marriage ridiculous in light of the idiotic racism that surrounds us here in South Africa.

I'm walking down Eloff Street; a line of African workers swing pickaxes into a roadside dig. They hold their rhythm, chanting. The upswing "*Abalungu*", the down swing "*goddamn goddamn*". White man, goddamn goddamn.

I've taken easily to advertising and already have a better job at another agency, Lindsay Smithers. I can afford to buy Basil's car. He no longer needs it. His current girlfriend owns a more desirable one; a convertible Chevrolet Power Glide.

Working alongside me at Smithers is Rod Dyer. Rod's an elegant young man three or four years younger than me. He has a *penchant* for American style, wears his hair crew cut, his trousers narrow, his tie slim. He admires my work but not the way I dress. He's curious about my politeness towards our African staff of cleaners and tea-makers.

We talk about it. I tell him about Bloke and my visits to Sophiatown an out of bounds area where Bloke lives. Rod admits he doesn't understand racism. Doesn't understand hostility towards people because of their colour. Brought up as a conventional White South African he's expected to adopt this unfriendly attitude as correct behaviour. He's not comfortable with it. Most of the American jazzmen he admires are Black. He'd like to know more about the Blacks of his own country. I suggest if I introduce him to Bloke and take him to Sophiatown, he can improve my sense of dress by introducing me to Spiro his tailor. It's a deal.

29.

Almost from the moment she arrives in South Africa Jean is depressed by the reality she had feared that now surrounds her. She's bitter. She interprets the slightest suggestion of an unkind word or action to an African as a racist slur. It angers her. She shouts it out and depression sets in. She's on her own. I'm embarrassed at the ill-mannered treatment of anyone of colour in the workplace but it doesn't dampen my enthusiasm for the work itself and I thrive. Jean, less sanguine, finds difficulty at all times in controlling her contempt for the privileged White racist bourgeoisie.

Her depression renders her negative about most things and difficult to live with. I find myself skirting around conversations that might further upset her. I'm concerned. I can't give up. Both of us struggle to find a place for her here, something less confrontational, teaching perhaps; she has two Arts degrees. She'll think about it. She comes with me to the Bantu Men's Social Centre to see if there's anything she could do there and as we enter there's a palpable change in her mood. She smiles for the first time in a week. Not only does she see Africans, she sees a room full of working class folk she can identify with. She's experienced their frustrations herself, their sadness and anger. She understands what it feels like to be on the wrong side of the tracks. It lightens her heart to be amongst them and her depression dissipates. I think she has found her people.

Everyone at the Centre lives and works in Johannesburg. They live in the Townships at the edge of the white City and come in everyday for work on overcrowded buses or trains. The wealth of Johannesburg attracts people and their families from every tribe in South Africa. They eke out livings as second class citizens serving an affluent society at very low pay doing the things the Whites don't want to do. Women mostly find work in service to White families.

Their tenancy in the City is precarious. One wrong move, an irregularity in a passbook and they can be sent back to the greater poverty of their homeland. Most of them, particularly the young, enjoy the jazzy, dangerous lifestyle of the Township. No one wants to go back to the farm.

One of my guitar students, a young Zulu woman, tells us she has no job, no place to stay; the Police are going to send her back to Zululand. She wants us to employ her as a servant then she would have a Pass and a place to live, the servant's quarters top of our apartment building, and avoid being expelled from the City. Jean tells her we don't need a servant but we'll stand in; she's welcome to occupy the rooms up there.

Having a servant isn't something we could be comfortable with. Everyone else has at least one. Servants are a point of gossip amongst middle-class White women forever criticising them, their abilities as cleaners or cooks or their untrustworthiness or their stupidity. Even down and out families, drunken refugees from farm poverty, the white trash of a wealthy urban society, manage to squeeze in a servant or two to satisfy a fading sense of race superiority.

30.

Bloke Modisane introduces us into Johannesburg's multi racial elite; our Socialist ideals lead us into radical left wing circles centred on the African National Congress. Bloke doesn't engage with politics, doesn't belong to a political party. He deals with apartheid every day in his own way, managing to conceal any bitterness he might feel behind sardonic wit. He introduces us to his editor at Drum, Sylvester Stein, a white South African ten years our senior who'd recently taken over editorship of the magazine from Anthony Sampson. He stands when I enter his office; we shake hands. He struggles amiably to place me in his own genealogy, forever trying to uncover distant blood connected relations. When he sits down I notice he's wearing two quite different socks. He talks nostalgically about Soho London and his failed attempts as an aspiring actor. We warm to each other and arrange a meeting of families.

Drum's a Black magazine with White roots, owned by Jim Bailey, son of Sir Abraham Bailey, 1st Baronet, KCMG, a South African diamond tycoon, politician, financier and cricketer. I meet Jim a few times at the magazine and, with Jean, at a couple of staff parties at his farm. During the war he served in the RAF as a Hurricane fighter pilot.

I find him offbeat and congenial but he doesn't seem to enjoy that kind of reputation with all those who work for him. His magazine purports to be a celebration of the non White community but to Jim it's a business that has to skirt around sensitive race issues in order to survive.

Jean and I become close friends with Sylvester, his English wife Jenny, their four kids and their dog, a black Labrador named Zumba. They're a post-bohemian bunch, no heavy-handed family rules or regulations. Not chaotic, just easygoing. Sylvester is part of an eccentric family of Jewish intellectuals; his elderly absent-minded dad has been a professor of mathematics at several universities.

Jenny is the daughter of Alan Hutt, a well-known English journalist, Marxist, social historian, chief sub-editor on the Daily Worker and an authority on newspaper design. She was a film editor for director Humphrey Jennings with the Crown Film Unit in England. When we meet she's an active member of Black Sash, the White South African women's voice against the apartheid regime.

She and Sylvester are a strange mix. By and large they get on well together. Sometimes there are storms when Jenny thinks he's sleeping with the wife of one or other of their friends but their sparky relationship endures and new chapters open. The Steins mark our arrival in their lives by opening a new family photo album entitled "Hart attack".

We hang out with anyone active in anti apartheid politics. We don't see ourselves as potential combatants in future plans but are drawn to the cause and want to help. There are several political organisations lined up against the government. The ANC, the most popular, represents the idea of a democratic multiracial society. That's where our hopes lie. The Pan-Africans want an exclusively African country. The Indian National Congress parallels the aims of the ANC. The predominantly White Congress Of Democrats and the major trades unions are mostly Marxist split by slivers of doctrinal interpretation with no one taking the lead.

31.

Much of our social life is spent hanging out, partying, with our new friends; the Slovos, the Goldreichs, the Bermans, the Blooms, the Sacks, young African lawyers Nelson Mandela, Walter Sisulu, Oliver Tambo.

Gradually we get to know the writers and photographers at Drum on a more personal level. Henry Nxumalo, Zeke Mphahlele, Lewis Nkosi, Can Themba, Todd Matshikiza, Arthur Miamani, Casey Motsisi, Gwi Gwi and photographers Peter Magubani and Bob Gosani under the tutelage of Jurgen Schadeburg a young photographer from Berlin. He's been at Drum since its inception; longer than anyone else and carries in his head a detailed inventory of its history. He remembers editors and sales staff and the naive but effective way Bailey evolved a sales pitch, mainly in the townships.

Friendships develop. Now and again we invite some of our new African friends to dinner when brutally frank conversations take place about apartheid, all of us searching for some psychological understanding of racism. A trust evolves between us; were all on the same side. When our guests are leaving a little drunk, to return to their homes in one or other of the townships I have to write individual notes asking whatever authority they might encounter to allow this man to pass. He is late because I kept him to do a job for me. Signed. Our friends did not have cars, they walked, some of them a tidy distance, plenty of time and space to be stopped and questioned by cops.

Many White radicals are Jewish and know what it is to be persecuted for no reason other than religious or cultural differences. Maybe we've evolved, over the centuries, a kind of genetic affinity with the experience of abuse. It would be difficult for a Jew to justify apartheid although many try and many manage successfully.

32.

The Bantu Education Act has just become law, excluding Africans from the standard education given to Whites; only a special government-designed education remains available to them, schooling aimed at cementing their servant status. Real education has become illegal. Without education Africans will be unable to communicate with the world around them; without being able to read freely they won't know what's going on, won't know what are their options in combating apartheid. Their interest in politics will be stifled.

The ANC innovates new methods of schooling. Nelson Mandela organises what he chooses to call Culture Clubs in the townships, centres of education that will evade the new law by avoiding the use of blackboards, exercise books and all other identifiable teaching aids. He knows us and our backgrounds and asks Jean and I to help evolve a system that allows teachers to teach children how to read and write without using these identifiably illegal facilities. The solution is obvious to us; we suggest painting as a means to learning vocabulary and the reading of it; painting pictures then painting words that describe the picture. Simple is as simple does but it's a beginning. We teach this formula to the teachers who will pass it on to the children.

33.

Living life both sides of the political/racial divide is stressful. It's Saturday night. Another party at the Stein's house to let off this week's tension, dissident friends, high on adrenalin and alcohol, hooting, hollering and dancing. The Stein kids in pyjamas ready for bed chase Zumba the dog from room to room. Behind the loud laughter and conversation, is the paranoid thought of a police raid; contingency plans are in place, a quick back door exit for Africans.

It's said these drunken nights of release are incomplete without our songs and we're happy to contribute them. One time we're absent from the party being early to bed on account of the previous night's drinking. We're wakened to find Sylvester, Nelson Mandela and Walter Sisulu, jigging around our bedroom singing *"Jikela emaweni sia hamba,"* a popular Manhattan Brothers song of the day. They insist we get out of bed and join the party.

For me, these get-togethers are a mix of pleasure overshadowed by my ever-present anxiousness of watching Jean dancing and enjoying herself with other men.

The most memorable of all parties is at Bloke's Sophiatown home. Sophiatown is an anomaly in the apartheid plan built at a time when Blacks were still allowed to own property. When non-Whites moved in, Whites moved out and the value of the property fell. I guess cash-poor property owners turned to renting every square inch of their space to pay off mortgages and survive. The town is crowded with dwellings. In back yards one or two dilapidated wood and corrugated iron constructions. It's never become a government-controlled location. There are no cops allowing you in or out. It sits there open to all, an obtrusive sore between two very White middle class suburbs.

Sophiatown stands as a symbol of anarchy in this tightly controlled world of apartheid. Bloke was born here, grew up here. He knows the layout and the smells. He knows the people of all ethnicities. He knows the cops and knows the villains. Sophiatown houses a raw, unruleable society prey to violent street gangs. Murder is commonplace. Police sorties frequently result in bloody fights with the locals, sometimes ending with a death or two.

Bloke occupies a room in a run-down house belonging to his mum, Ma Bloke, a brewer of illicit *Skokiaan*. His door lets out onto a scruffy yard of broken stones and refuse. His room is small; a bed more or less fills one side of it, next to it a refrigerator, a Decca portable record player on a small coffee table standing close by and one semi-upholstered chair. A large poster of John Wayne for the film *Red River* covers the wall above the refrigerator and by the small curtained window are pinned postcard reproductions of paintings by Paul Klee, Cezanne and Manet.

He didn't mean to have a party. Jean and I are there with John Goldblatt and a bottle of illegal brandy listening to some Bach on the record player. Whenever our car is seen parked outside the rusted corrugated iron fence surrounding the property, it attracts curious passers by to Bloke's door. Everyone in Sophiatown knows him. We're laying back listening to the music, two or three of them knock at the door; Bloke opens and they invite themselves in. Within a few minutes, another knock on the door. The little room is getting crowded.

The locals shyly acknowledge us, Bloke's White visitors. When they have sampled the brandy they become more relaxed, want to get up and dance. The record player is still beating out Bach's *Concerto For Three Harpsichords*. His guests query Bloke's taste in music. "What is this noise?" they ask, and insist he replaces it with something African they can dance to and he does and they dance, we all dance. It's not ballroom dancing it's called *Kwela;* it doesn't take up much room; the form of the dance is largely rhythmic movement of the body parts. Everyone admires Jean's *Kwela*; they say she dances like an African girl.

34.

Above the noise of this illegal, lively and slightly drunken party, another knock on the door. Bloke opens it to an African policeman an *assegai* in one hand, a young man handcuffed to his other wrist. Bloke backs up and allows them into the room. The cop sniffs the air and spots the brandy. In order to mitigate his guilt for breaking the law about alcohol, Bloke offers him a drink, which he accepts, sips then asks for a cigarette. This cop has an air of sly authority about him. He has a roomful of law-breakers at a disadvantage. Space is made for him to sit, cigarette between lips, brandy in hand, his manacled prisoner on the floor beside him.

Jean tries to charm him, to get him to release his prisoner but he refuses, indignant. He'd just caught him breaking into someone's house and arrested him. He shakes his head,

"No madam. He's a prisoner".

After another drink he stands dragging his prisoner to his feet.

Manacled to each other, they dance to the music.

Black South African music fascinates us. Bloke takes us to events at the Odin, an old rundown cinema in Sophiatown that serves as a venue for music and other social activities like political meetings and such. We're frequently there and don't feel out of place. We get to know the famous singing quartet the Manhattan Brothers and their soloist Miriam Makeba, the talented trumpet player Hugh Masekela, the guy who, when he was still a kid, was given a trumpet by Trevor Huddleston which had been donated to his school by Louis Armstrong. Phenomenally modern jazz pianist Sol Klauster; growling tenor sax Mackay Davashe and legendary alto sax Kippy Moketsi. We didn't get to know them beyond their music but we listened to them a lot.

35.

I change my job again for more responsibility and better wages. I work now for J Walter Thompson. I think Jean is having an affair with Rufus Khoza, one of the Manhattan Brothers. I'm not sure but stung by the thought. The mind dwells on it; supposes African men to regard the sexual conquest of a White woman as some kind of trophy attesting to their manhood, their self respect, their daring. Jean spends much of her time with them. This weekend she's going down to Durban to a concert with her friend, the popular singer Dolly Rathebe. Jean adds, idly, Rufus is getting a lift with them.

I'd like to be completely free of these nagging doubts but the mind won't allow. It dredges up the same old pain of possibility. Paranoia. It's there again. Is it any different now we're husband and wife? It can't go on like this but it does. Distance is growing between us. I can't accept either

her word or her complete independence. It's not the way it's supposed to be.

Our own music keeps us together. When we play and sing there is honesty and trust between us; no room there for any bullshit.

Our friendship with Sylvester drifts into his professional life; sometimes thinking up offbeat stories that might prove meaningful enough for him to publish. Jean and I know a few popular African songs and Sylvester, looking to fill a couple of pages, suggests we take our repertoire into Sophiatown and start singing in the street.

He sends his photographers to cover our performance and the crowd of curious locals that gather around us. They'd not heard White people singing their African songs. Not everyone is happy about our presence there. Some shout we're there to distract people while *Tsotsis* burgle their houses.

The managing director of JWT where I work sees the pictures in Drum and is not happy about them either. I hadn't given him a thought. He wouldn't know anything about Drum. Few Whites did. It was mainly distributed through small shops in the townships. He hauls me into his office, says had I been anyone other than just a 'back room artist' he would have fired me.

36.

J Walter Thompson was the first advertising agency here to employ an African in an executive position. Nimrod Mkele, a handsome forty year-old Xhosa with a degree in psychology is in charge of research into the African market. Few people in the agency acknowledge this or, perhaps, don't even know of his presence. I was impressed. I'd never heard of an African being raised to the level of major company executive. Did this herald some kind of change in White attitudes? It was a thought.

JWT was thinking ahead. Africans are more than ninety per cent of the population; they will be the market when they've sorted out the politics. That's sure recognition that apartheid has, in the minds of commerce and industry, a limited life expectancy.

37.

Sylvester decides to do a Drum story by himself, the kind of thing for which his star reporter Henry Xnumalo is famous, placing himself in the action, Henry getting arrested to get an inside prison story. Sylvester sits while Jean blackens his face and neck down to his shoulders, hands and forearms with a solution of potash of permanganate. He sits like the actor he might have been being made up for his performance. She dyes his brown hair as black as an Indian. He plans to go and live for a week with one of his staff as a Coloured man and perhaps arrive at a deeper insight into non-White life.

The adventure breaks down. Within two days the Permanganate begins to wear off leaving pale blotches on his face and hands. He needs to be anonymous but has become noticeable. I don't think he wrote it up.

38.

Sylvester sometimes turns to us to do things that only a White person could do with impunity. He wants us to con our way into the Fort, Johannesburg's main prison, and convince the Governor to let us see Mary Louise Hooper, a wealthy American heiress, civil rights and anti-apartheid activist and the first white member of the African National Congress. She's being held awaiting deportation. He wants to do a piece on her in his next issue and needs a dramatic picture of her behind bars.

Sylvester's Leica in my pocket, we approach the huge and forbidding doors of the Fort pretending to be the closest of friends of someone we don't know, not even what she looks like. I ring the bell. A prison officer appears and asks what we want. We want to see the Governor. He picks up a telephone, has a brief conversation in Afrikaans. He nods to us. We're led to the Governor's office. He greets us politely. We tell him his prisoner Mary Louise Hooper is a dear friend and we would like to say farewell to her before she's deported. He mumbles and grumbles and finally decides to let us see her. He apologises for his hesitancy, says he hears so many cock and bull stories.

39.

We're led into one side of an open quadrangle fenced all around with chicken wire and left by ourselves. After a few moments Mary Louise Hooper whom we'd never in our lives seen before is lead into the other side of the yard attended by a wardress. Mary Louise stares at us through the wire trying to figure out who we are. We manage a smile and a feeble wave of the hand. What do we do now? I couldn't drag the camera from my pocket while the wardress was there.

Like you might only expect in a movie, the office telephone rings and the wardress hurries inside to answer it. Out comes the camera. I only have time to get off one frame of Mary Louise, slightly out of focus, looking incredibly bewildered, not behind bars, behind chicken wire.

Word gets around the ANC network fast. As we're leaving the

Fort, a young African gets out of a car parked across the road. He approaches us. Says Chief Luthuli would like a word; he's in the car. Elderly, handsome, Albert Luthuli, is head of the ANC. He wants to know if his good friend and collaborator Mary Louise is alive and well.

40.

Things are coming to a head. The bus company that plies daily between the townships and the city, packed to the gills with workers, decides to raise the price of a ticket. There's unrest and protest in the townships. The passengers have, as it were, gone on strike and are walking the six or seven miles to work in the morning and then back to the townships at night. The Black Sash women, although small in number relative to the task, drive their cars back and forth between city and township, lifting as many of the walkers as they can.

It's 1956. International reaction is encouraging the non-White community in its protest, voices are beginning to be heard, public meetings convened. The police are edgy and unpredictable. Sylvester suggests I keep a camera with me at all times. He'll call me at my workplace if anything significant occurs.

The government has arrested a hundred and fifty six activists including Nelson Mandela and other members of ANC, accusing them of treason, a move calculated to impress upon the entire nation the government's determination to stifle all opposition to the *status quo*.

Drum is a hive of activity, putting together double-page spreads of photographs of each and every defendant. Sylvester calls me at work. He says there's a huge demonstration outside the Drill Hall where the trial is being held.

I run across town to the Drill Hall with camera arriving in time to photograph Police, a lot of them young and frightened, inexperienced hands on butts of pistols, trying to disperse the crowds. The noise is intense as the African crowd resists, their shouting and ululation over scored by the whining sirens of Police vehicles. I'm unsure what is happening. There's an air of apprehension, a mix of excitement and fear, something is changing before my eyes; perhaps I'm seeing the emergence of a new Black confidence. Although a socially close friend of the revolutionaries I have no idea of their militant strategy, the encouragement of Africans to dissent and mobilise against apartheid.

41.

Jean relates to Africans in a very personal way, her attack on apartheid is personal. She despises and ignores its laws. There's no doubting her genuine love and respect for people under the cosh. Media is my playground, but bringing the evils of this regime to the attention of those who could make a difference is proving difficult. I have no success. There are those, here and overseas, that could bring pressure to bear on the South African government but each has their reasons for not wanting to do so. Prime Ministers of some European countries refer to Nelson and those around him as terrorists beyond the pail.

Subtle personality differences, marginal differences in point of view between Jean and I are beginning to tell in our domestic life together. We argue loudly about things over which we have no control. Neither of us is happy. We consider having children thinking they might cement our relationship but no soul seeking to emerge into the world wants to take the chance. We still share some activities, we go to meetings together, to parties, we sing, but the length of time we spend by ourselves is diminishing. I don't know what to do about it. We're getting lost.

42.

Part of our personal campaign against apartheid is to help Ian Bernhardt, a serious minded, ambitious show business producer, to organise African Jazz concerts to be performed for the first time ever to White audiences in the Selborne Hall, the municipal heart of Johannesburg.

Crowded in our apartment we are auditioning musicians and singers. We're listening to their songs, making suggestions about presentation. Being observed by White neighbours across the street we draw the curtains. They made headlines of it the following day in Dagbreek, an Afrikaans daily tabloid. SKOKENDE BOND PARTY IN DE GOUSTAAT. (Shocking mixed party in the golden city). Nothing comes of it except a warning to ourselves to be more circumspect.

I design posters for the event. We have them printed and distributed. The concert fills Selborne Hall for three nights. The quality and integrity of the performances erases, at least for a moment, the veneer of disrespect that Whites normally show to anyone of colour. They clap loudly.

This new creative energy flowers into the planning by Ian of a full length African musical about the life and times of a legendary, talented heavyweight boxer, Ezekiel Dlamini, known to his admirers as "King Kong". After a meteoric rise in the world of boxing, Dlamini can't handle the fame and it's downhill all the way into drunkenness and gang violence. He knifes his girlfriend. In utter remorse, during his trial he asks for the death sentence and gets 14 years hard labour instead. After his release from prison this year he was found drowned; it's believed his death was suicide. He was thirty-six. It's a story of operatic proportions. Harry Bloom has committed to writing the book. Todd Matshikiza is already humming the tunes. He and Pat Williams are thinking about script and lyrics.

43.

Bloke and his girlfriend Ficky are getting married. Jean and I and Goldblatt are invited to the wedding reception at her home. Ficky, a pretty, personable young woman, is the daughter of a respected court translator of native languages. She and her family live in Germiston Township; at the checkpoint Bloke has to plead with the cops to allow in his White guests. They grudgingly allow but insist one of their own attends the reception to see that nothing even more illegal takes place. At the reception, when Bloke produces a bottle of brandy the cop falters. It's against the law for Africans to drink brandy. He could arrest them. Jean charms him into forgetting the rules for an evening. After a few sips of brandy, he loosens his tie, unbuckles his gun belt.

Our constant friendship with Goldblatt is under threat. His friend, Faye, from England, the one I remembered seeing him with in the St Martin's common room, has arrived and is staying with him. A strange young woman, she invites us over for a drink; John's not there. Is he coming? She hunches her shoulders as if protecting a secret. She pours wine and we drink and chat about the Charing Cross Road. She wanders to the window. Turning to us eyes lowered, she touches her lips with her fingertips in a gesture of reticence and starts quizzing us about John. What did we really think of him? Neither Jean nor I have ever considered John in a critical light. We're baffled. We have nothing to say. Where is he? As we leave, we see him, or think it's him, a bulky body lying outside under the window concealed in a sleeping bag. I guess he'd been listening. Strange, eh? I'd never thought of him to be sensitive to what others thought of him. Whatever, our friendship is fading.

Even though we're co conspirators in a strange country, Jean's casual relationships, real or suspected, continue to cause me anguish, anger and jealousy. It's December 1957 when she leaves me. This time she's moved out and there's no longer a question of reconciliation. Flying in the face of the law, she's living dangerously somewhere in a White Johannesburg suburb with our friend Can Themba. That's what the Special Branch guy wanted to know. Did I know where she was living? I lied. Sure I knew.

I'm hurting, my ego bent out of shape, not knowing how to deal with this grand social embarrassment. With Jean gone, my life is in limbo. I struggle with it for a while. Sylvester's wife Jenny says I should have dumped her years ago. Bloke, without taking sides, is supportive; he introduces me to a sweet young Basuto lady, a friend of his named Margaret; he calls her Princess. He makes his small home available to us whenever we want to keep company. She works in a Rothmans's cigarette factory and whenever we meet brings me a carton of cigarettes. Apart from the sweetness of Princess's undemanding friendship, life has become dull and less tolerable.

At work they know about Jean's departure and assure me it's the best possible thing that could have happened. She's a troublemaker. Whenever invited to one of JWT's parties all she does is argue impatiently, often angrily with important clients about apartheid. They thought I'd be better off on my own. I find myself constantly readjusting to Jean's absence and the ignominy of the cuckold. But her departure clears my mind of the uncertainties. My sadness is slowly overcome by a seductive sense of freedom from the constant anxiety I felt in her presence, I can see possibilities and avenues I'd never thought of opening up.

44.

Sylvester's finished his satirical indictment of White bourgeois attitudes to apartheid, titled *2nd Class Taxi*, I design its dust jacket. If it ever gets published it will surely be banned here. Black Beauty, the 1877 novel by Anna Sewell about animal welfare, was banned not only on account of its seditious title but because the book was largely about how to treat people with kindness, sympathy, and respect.

I'm thinking of leaving. Some of our friends both Black and White have already left. Es'kia Mphahlele and his wife Rebecca managed to get passports to allow him to take up a teaching position in Nigeria, Lewis Nkosi is in the United States teaching English Literature at Wyoming University, Bloke's had offers from several American institutions but is hanging on. I don't think academia has any great appeal for him.

Sylvester resigns his editorship of Drum over an editorial difference with his boss Jim Bailey. I guess it was his last gesture of defiance. Sylvester wants to run a cover picture of Wimbledon tennis champion Black Althea Gibson embracing her defeated White rival Darlene Hard. Jim doesn't think it a good idea; looks like trouble. The government would surely shut them down. Sylvester resigns. He and his family prepare to move to England.

I learn the UK government is raising eligibility for conscription from twenty-seven to thirty-one. Another four years? I don't think I can make it. Danger is all around; the Special Branch visiting me without appointment; my wife living with an African in the middle of town. It's time to go. I can't take another four years of this. I'll take my chances with conscription and join the exodus, turn my back on a failed marriage and the insanity of the apartheid insult and return to England.

My family had grown since Audrey, an artist in an advertising agency, moved in next door. She fell in love with Basil. He recognised her as the girl he'd been waiting for; they soon married and are now expecting the birth of their first child. I'm not happy about leaving them all again but I have to get out of South Africa.

45.

I've been living in Africa four years and seen little of it; a week in Cape Town, another in Durban, a weekend in Lourenço Marques with Jenny and Sylvester, a few days rock climbing in the Magaliesberg with Frieda and a Christmas in the Kruger National Park learning how to keep champagne cold in thirty degree heat.

As a penniless student, when I wanted to go to some place out of town I'd hitchhike, sometimes with Jean. I'd hitched up and down England and, every long summer vacation, to Paris and other places across Europe. There's something inherently exciting about hitchhiking, meeting people in a mood of generosity, people you might otherwise not know, getting to know them for the duration of a journey never to see them again. A lottery every time a car or truck comes down the road. The excitement, the exultation when someone slows down and stops to pick you up.

I'm going to hitchhike back to London and see something more of this great African continent. Anthony Smith did Cape to Cairo last year on a motorbike but I've not heard of anyone hitchhiking.

In the best possible of worlds I would head north to Nairobi, then on to Juba in the Sudan, the first navigable point on the Nile, then by boat to Alexandria. From there a boat to Europe, Greece or Italy, then out on the road again direction London; an exciting prospect in the best possible of worlds.

But the chances of being able to hitchhike all the way back to England had been diminished when Egypt decided to nationalise the Suez Canal. After a military debacle and ignominious withdrawal of the British, French and Israeli armies, the Egyptian government cut off diplomatic relations with all of them and, ever since, refused visas to any of their enemies' citizens. If the diplomatic situation doesn't improve I'll have to figure out another route. It's never that easy; there's always something.

46.

I love Jean and probably always will. I wish we were still together even knowing her loyalty could never be relied upon. I'm sad at the finality of it all. We've been partners for more than seven years and now I'm thinking I'll probably never see her again. I'm worried her reckless affair with Can could end disastrously not only for them but for my family. It's on my mind. I call her; she agrees to meet.

We're sitting drinking coffee at a sidewalk café in the sun, like strangers thinking they had once known each other. I tell her I'm going home. I don't like it here. She nods sympathetically. Silence while she plays with her sunglasses. She asks would I be flying back. I tell her I'm going to hitchhike. She takes it in; a few more moments of silence. Quite casually, she says she's been thinking about going back herself. There's nothing for her here in South Africa. She'd tried to get Can to come with her to London; there are ways and means; he would be free and their relationship no longer subject to ridiculous race laws. But he didn't want to leave, felt his place, come what may, to be in Sophiatown. More silence. She asks if she can come with me.

47.

Jean moves back. It's not like old times. Perhaps it never will be. Some events in a relationship can be forgiven but never absorbed and forgotten. We have the trip to consider, the gear, what equipment would be useful on the journey. It preoccupies us and shifts our focus from recent events to the present. We'll need a map, certainly.

In a map shop we find a hefty, three hundred-page book called The Highways and Byways of Africa. It references every road on the continent, major and minor, even Hippo paths in the North-eastern deserts of Kenya where we might be headed. It would be heavy baggage but worth its weight. We're trying to figure out what we might encounter, what we might need in different climatic conditions and local circumstances. We acquire far too heavy duty army rucksacks, water bottles, suitable khaki clothing, something warm, desert boots, sleeping bags, a mosquito net, a sharp hunting knife all from an ex army store. Besides these essentials, a camera, a guitar and a snakebite kit from the Johannesburg zoo.

We pack the rest of our lives, clothes, books, records, photographs, my advertising portfolio, into a couple of tea chests and ship them to cousin Peter's flat in England, the flat where he and I had lived together and where Jean and I will stay if and when we arrive back.

We spend our last evening in Johannesburg eating dinner and getting drunk with Lionel and Ellie Rogosin. Bloke introduced us to them. Anyone arriving in Johannesburg in need of help or advice would somehow find Bloke. Lionel is friendly and positive. He's a young filmmaker from New York with a reputation for wanting to know and tell the truth. For purposes of Government permissions he's pretending to be in South Africa making a tourist documentary while, in fact, he's making an epic exposé of apartheid. He questions us wanting to know what it's been like living here. There's much to talk about. He's already inside the film he's going to make. There's a script that tells of the experiences of a young African adult who comes to Johannesburg from a farm in Zululand to find a job and a place to live. Lionel has planned talk sessions around this unsophisticated man in company with city-wise Bloke, Can Themba and Arthur Miami to be filmed over a drink at *The Thirty-nine Steps*, a Sophiatown *shebeen*. The script allows these three well-informed journalists to talk freely about their thoughts and expectations of apartheid. Seditious stuff. He's getting Miriam Makeba to drop by and sing a song to liven up the scene. We admire Lionel's critical attention to apartheid.

48.

When Jean went to live with Can, I saw a lot of Felicity, an older woman bored with her wealthy husband and soft on younger men. We became friends. She offers to drop Jean and I at the start point of our journey, the highroad North outside Pretoria.

It's ten in the morning when we get there, hung over from the evening with the Rogosins. Felicity helps unload our cumbersome baggage onto the roadside. She glances at Jean then smiles briefly at me. Wishing us both a safe journey, she turns the car around and drives off.

The Great North Road doesn't stream with traffic. Every thirty minutes a car or two, a truck or two, mostly local, pass us by. It's near midday before someone stops for us, a genial young man in a Chevrolet. Jean sits in front, I get in the back with all our gear. The driver introduces himself. Dirk. He's a fertiliser salesman. He's a talker.

"Where you headed?".

Jean replies,

"London. But Bulawayo will do".

He laughs, shakes his head, starts on a long and uninteresting story. Jean, smiling, pays rapt attention to his boring chatter. In the back, I'm getting those feelings that creep over me every time she shows an interest in another man. Dirk says he can take us as far as Beit Bridge, the border crossing over the Limpopo, the river that divides South Africa from Southern Rhodesia. He'll be turning east the other side of the river into the farmlands of Mozambique.

The great, grey, greasy Limpopo looks as bad as it sounds. We stop here for an hour. At Beit Bridge the Limpopo is at low elevation, hot and humid beyond comfort. Dirk makes this journey frequently and is completely at ease with its atmospheric density; he takes us to a café bar by the river and buys us beer. It's late afternoon and the insect life is intense. Bugs, big as golf balls, buzz and flutter around our sweaty faces. Dirk's amused. He picks them out of his beer and drinks on. He swats a flying beetle away from Jean's mouth.

Across the border, after passport control, the two lane metal road divides into two tarmac strips little wider than the breadth of a vehicle's wheel. If there's oncoming traffic, both vehicles swing over, each keeping two wheels on a tarmac strip. Dirk maintains his stream of chatter. We reach his East bound road. It's early evening when we say our farewells. I'm happy to have him out of my life. We watch him drive away down the dirt road disappearing in a cloud of dust and evening gloom leaving us in the middle of nowhere. Moments of apprehension.

It's growing dark. This is our first night out in the *Bundu*. In the fading light Jean looks around the empty landscape her sense of adventure momentarily diminished. I check in the Map Book but find no town or village or hamlet of any kind in walking distance where we might lay up for the night; chances of another lift are doubtful. We move our baggage a hundred yards or so from the road then look for and find a comfortable conformation of the ground to sleep in. We bring our baggage to this spot, unpack the sleeping bags, unroll them and sit down.

49.

Jean gives in to the situation, shrugs, gets out our food; cheese sandwiches, biltong and tinned peaches. It isn't much but it'll do. Long distance hitchhiking is much about improvisation, adapting to new situations. The air is warm and free of insects. We take off our boots, zip the sleeping bags together and fall asleep in each other's arms.

Sunrise. Jean, half awake, takes a moment or two to remember where she is and why. I'm already up, consulting the Map Book. After shaking out our boots we put them on, breakfast on the remains of the tinned peaches, wash teeth with water from the water bottle, roll up the sleeping bags, pack them and drag the gear to the roadside. It's eight o'clock, already warming up and no traffic.

Time passes slowly. It's midday before someone stops; an elderly fellow, an English farmer in a Land Rover. He upbraids me gruffly.

"What kind of man brings his lady wife out on the road like this? You should be ashamed of yourself! I wouldn't have stopped but for the little woman."

He can say what he likes; we're on our way to Salisbury. We travel in silence. He lets us out without farewell when he needs to turn west off the road to his farm.

Waiting at the roadside in the heat of the afternoon. As if from nowhere, a young African lad carrying an old dining room chair trots out of the bush. We ask what he wants. He doesn't speak English. Pointing to the sun and then to a tree, he leads Jean under its shade, sets down the chair, sits her on it then disappears again.

Within the hour we have another lift taking us via the Zimbabwe Ruins all the way to Salisbury. We pause at the ruins, the remains of brick towers and enclosures built more than five hundred years ago by a prosperous, well-organised society long before the coming of the White man.

50.

It's dark when we arrive in Salisbury. We'd heard of a multiracial hotel, first of its kind anywhere in Southern Africa. We find it and check in, catch breath, revive spirits, take a shower and find a place to eat. That night we sleep in a bed like old times. Late the following morning we repack the rucksacks, check out of the hotel, wander around the town, find a café and order breakfast.

A young man drinking coffee sitting close by, intrigued by two heavily equipped hitchhikers one of them an attractive young woman, comes over to our table and introduces himself. He's a producer for Salisbury Radio; he thought we looked interesting enough to interview. After breakfast we go with him to the radio station just round the corner. The interview is short, our host wondering at our courage or stupidity to undertake such a journey; we joke; he laughs, comments on the guitar, asks if we'd give his listeners a song. We sing. He enjoys it. We enjoy it. Then back on the road looking for a lift to Lusaka, capital city of Northern Rhodesia.

During the years we spent in South Africa, politics had become central to our lives, our major source of interest. Before leaving Johannesburg Sylvester had given us names of people to contact, journalists with an ear to the ground, a finger on the pulse. He liked to keep in touch, find out what's happening in remote parts.

51.

Cyril Dunn is the Africa correspondent in Lusaka for the Observer. We meet with him. He's middle aged, friendly and a mine of information. He's covering Kenneth Kaunda's progress towards independence for the country whose name he will change to Zambia. At the present speed of decolonisation Kaunda will soon be its President. In his pursuit of news Cyril had become friendly with him. Would we like to meet him?

Evening. Cyril's driving us to an African Township to the north of the city; concrete houses in serried ranks. Kaunda's is a one-room breezeblock affair common to most of the cheap housing White city councils provide for their Black communities.

It's dark when we arrive. Calm and friendly, Kenneth Kaunda, soon to be President, invites us into his one room home. His wife and children are asleep on the concrete floor in a corner; he works at a small desk the other side of the room on a typewriter by the light of a paraffin lamp.

It's a special moment, being in the presence of a man about to make real his political dream, to create an African State out of this old White British colony donated to the Crown a hundred years ago by Cecil Rhodes. He's politely curious about who we are, what we do and where we come from. There's little talk of political interest. What can a couple of English youngsters, running away from it all, possibly say to him? After ten or fifteen minutes, wishing him success, shaking hands we say goodbye.

52.

Before we leave Lusaka, Cyril tells us of an eccentric colonial Englishman, 'Cherapula' Stevenson, who, in early days of colonialism, had gone native and fathered with the help of many African wives a considerable tribe of children. He thought we might be interested to meet him, a genuine relic of colonialism. He lives not far off the road we'll be taking.

We follow his directions but arrive too late. Cherapula died a few months ago. We visit his grave, a large earthen mound draped with a tattered old Union Jack. The day of the Whiteman in Africa is coming to en end. The sun is setting on another British Colony. We're back on the road heading for Mbeya a thousand miles away across the border in

Tanganyika; a large town by African standards.

53.

The smooth tarmac strips come to an end, leaving us on a road of graded red earth. It's bumpy and hilly but with magnificent views of cliffs, rivers, plains dotted here and there with Baobab trees. It takes us four days, nights and numerous lifts to get to Mbeya. We spend the nights camping totally alone under the vastness of the Milky Way listening to unrecognisable grunts, groans, coughs and moans of unseen animals. This is the Africa that's been hiding from me these four years.

In Mbeya we report to the District Officer. We're expected to report at each destination to prove we're not stranded. On the road again, we find a lift to take us towards the next major coordinate on our map, Iringa, a small town about three hundred miles to the north-east on the road to Dar Es Salaam, our vague destination. It's late afternoon on Christmas Eve when we are dropped off a hundred miles short of Iringa at a motel in Sao Hill. We didn't expect to find a motel in the middle of the African bush. It must be quite new; it's not mentioned in the book. In our present state of sweat and exhaustion it couldn't have been more welcome and we'd get to sleep in a bed for a night.

We walk wearily down its long drive. We're tired and dishevelled, badly in need of a bath, a meal and a real bed. A motley bunch of White people, drinks in hand, saunter out of the motel to watch our arrival. They're dressed for a party, the women in yesteryear's debutant dresses, men in dinner jackets, one of them wearing a kilt. They're curious about us. The man in the kilt, about forty or fifty years old, limps closer supported by a walking stick in one hand, holding a drink in the other to tell us in upper class voice the motel is closed for a private party and that we would have to get back in our car and drive on to Iringa. When we tell him we have no car, we're hitchhiking, he blinks. He's thinking.

It's growing dark. He knows the chance of a hundred mile lift to Iringa on this darkening Christmas Eve is very slim. He points to the guitar with his stick, says we can stay the night but we'd have to sing for our supper. He leads us into the motel followed by his amused and slightly tipsy friends.

We shower; dig into rucksacks for clean khakis. Bringing the guitar we join our hosts in the bar. The kilted Scot, Captain Stewart Stuart and his attractive, authoritative wife Emma own the motel. They cater for the occasional tourist and Commercial Traveller. Stuart looks rugged enough behind a full moustache, receding hair and slight limp, but his general demeanour, an almost obsequious deference to his wife and her wishes, show a less than dominant male. He'd served in the Army in India until its separation from British rule. Ten years ago, in 1947, he and Emma moved across to Tanganyika looking for the kind of pampered life they'd previously enjoyed in India. Now, ten years later, they're facing the possibility of their comfort being stolen from them again as Tanganyika prepares for independence. Beneath the Christmas levity we feel their resentment at the way things are working out for them, the decolonization of another home.

54.

Their friends and guests are all from a colonial class of which Jean and I are totally unfamiliar. I think we're as much of a mystery to them. A couple of local Eton educated farmers and their wives, a young newly wed couple from noble Knightsbridge families on their honeymoon, a thirty year-old, extraordinarily handsome Teddy Lardner, White Hunter and sometime Mayfair croupier, and Lady Barbara Warncliffe an attractive near forty year-old settler.

We eat their traditional Christmas Dinner then, over cognac and cigars, we sing to them. By now the atmosphere has warmed and our hosts and their guests are taking a much more friendly interest in these two strange ones who have come to them out of the night on a Christmas eve. We are as strange to them as they are to us. We sing our songs and they applaud and think we're absolutely splendid. As the guests depart, Lady Barbara says we must come to tea one of these days before we go on our way.

The following morning there's no longer a question of us going on our way. We are now part of the gang and required to stay at least for another day to beef up the Boxing Day soccer team.

Boxing Day came and went, a *shenzi* affair. *Shenzi* is a Swahili word to which we were to grow accustomed; it means uncivilised, uncultured, uncouth. Nevertheless we're now regarded at the motel as 'family' and can stay as long as we like. A few days later we take up Lady Barbara's invitation to tea.

She and her husband live a mile or so from the motel in an extensive mud and wattle palace. During the war David had been a Lieutenant in the 8th King's Royal Irish Hussars, a light tank regiment in the North African desert. He'd sustained a neck wound and had been sent down to Sao Hill for rest and recuperation. He became fond of the place; built this palatial if primitively constructed bungalow. When the war ended he returned to his Kensington home, packed it up and brought it back here, and furnished this bungalow, where we are taking tea, in the style of his former home. Some of the furniture and the frames of large family portraits had been chipped and damaged in transit but still recalled, even in their new mud and wattle surroundings, times past of dusty elegance. Lady Barbara's eccentric husband David Ricardo is a grandson of the famous nineteenth century economist and close friend of George V.

Once married and settled David's ambition had been to become a cattle rancher in a land where the *Masai* owned all the cattle and only sold to each other. To buy cattle from them he had to become a blood brother, only then would they do business with him.

Having built a sizeable herd, in order to slaughter for sale to a largely Muslim community, he had to become a Muslim. Lady Barbara said she finds it infinitely preferable to his Masai days when he would rub mud in his hair and go out at nights howling at the moon with his new friends and business associates.

David joins us for tea smiling like Buddha, a Muslim-style woven cap on his head. He greets us briefly then seats himself on the carpeted floor, legs folded beneath him. He eats a biscuit, drinks his cup of tea, smiles at us for a moment, stands and, still smiling, leaves without another word. Lady Barbara looks at us and shrugs.

55.

There's an air of expectancy at the motel; they're waiting for something to happen. Emma's in the kitchen shouting at the servants as they prepare lunch. In the sun lounge, Teddy, immaculate in white Palm Beach suit, Old Etonian tie and white mosquito boots amuses himself reading a book about the history of tiger hunting in India and occasionally reading aloud to us passages about tiger hunting with small cannon strapped to the backs of elephants. Stuart is sitting in a rattan armchair cleaning a .470 calibre double-barrelled hunting rifle, a monstrous piece of artillery he'd brought with him from India. He and Teddy break into chatter for a few minutes about the general efficacy of various hunting weapons, become bored and slump back into fidgety expectation.

They're awaiting the arrival of the Mattheisen brothers, two White Hunters driving down from Eldoret in Northern Kenya. Jack and Mat Mattheisen. Teddy and Stuart are going to join them on safari. This year crocodiles crowding Tanganyika's rivers needed culling and have been declared vermin. No special hunting license required and crocodile skins fetch a fair price. I'm intrigued.

The Mattheisens arrive in a rickety old Bedford truck, a drop side with a canvas covering. They apologise for being late. They'd had to stop on the way to redeem their rifles from an Indian *duka* west of Nairobi where they'd left them in hock after celebrating the success of a previous safari.

Jack, about fifty years old, naked to the waist, wearing shorts, tennis shoes, an old felt hat probably discarded by his wife, is a rugged, deeply suntanned man of medium height, with twinkling blue eyes and a smile. Mat, two or three years younger, wearing shirt and trousers, shares his brother's blue eyes but little else. He's slight of build and his skin pallid, a congenial guy but of few words and much coughing.

Teddy and Stuart and Emma welcome them enthusiastically. Stewart Stuart introduces us like old friends and we're soon around the table drinking and toasting the up coming safari. They have a month, until the rains, to gather as many skins as they can. Jack, noticing my interest, raises his glass to me and invites me to come with them. There's no time to think about this kind of invitation.

By nature, I'm averse to any kind of killing. Crocodiles are not my favourite animals. They threaten, kill and devour unwary animals and human beings. They scare me. I rustle up some courage. Suddenly in my mind I see it is an opportunity not to be missed, an unexpected turn of fate. Crocodile hunting? Ugh. I hear myself accepting Jack's invitation.

Jean's more excited by the idea than I am, asks if she can come along. Jack, apologetic, shakes his head. No alcohol. No women. Either could give rise to mayhem among men with guns. Jean's disappointed and not relishing the thought of being sequestered for a month in the motel with Emma.

56.

In a meadow at the rear of the motel, Teddy jams a penny between two lower branches of a tree then retires about fifty or so paces to a deckchair. Slumped in the deckchair he takes aim with his 6.5 Manlicher and fires at the penny. After each miss he makes an adjustment to his rear sight, firing again until he knocks the penny out of the tree.

A large sack of rice, our food for the trip to be augmented by anything we could kill, is stashed in the Bedford along with an even larger sack of salt for preserving the freshness of the crocodile skins. Ammunition is prepared. Bullets are expensive. Apart from Stuart's double barrelled .470 both Mattheisens used the same calibre as Teddy. There were one or two unopened boxes for the 6.5 Manlichers and a bag of mixed cartridges that didn't quite fit in the chamber and required cotton thread to be wound tightly around the rim of their base. It seemed to me unprofessional if not dangerous but what did I know about guns and bullets. Stuart had ammunition he'd brought from India for his large calibre monster. It was old. He was unsure whether or not it would fire. The last essential we would carry with us was tobacco, really rough stuff cut from the stem ends of leaves. Each of us carried our own in an empty cartridge box along with a wad of papers cut from an airmail edition of a Danish newspaper that had found its way into the motel along with a stack of colour magazines.

Two young African men from the motel staff will come with us to handle heavy stuff, care for any campsites we might inhabit, do some cooking, things like that. My new friends did not show the same politeness for our Africans as they would for Whites and were brusque in giving them orders. I asked what were their names. Stuart said we call them Horace and Petrus. He didn't know their real names given them by their families. Discrimination by colour or culture never sits comfortably with me but after four years in South Africa I was sufficiently experienced to be able to deal with the chauvinism.

Stuart owned a battered World War Two American Jeep. He, Teddy and I would ride the Jeep; Jack and Matt would ride the Bedford with Horace, Petrus and our supplies in the back. Emma and Jean came out to see us off. Jean is not happy as we move off headed for the Great Ruaha River twenty miles or so to the north of Sao Hill. She shrugs and walks back to the motel.

57.

There were no roads. We lumbered our way across country, through bush and across stream. Our first stop and where we would spend the night was to be a Mission run by Padre Bianchi, an Italian Catholic order. Within a few hours the Jeep's clutch plate gave up the ghost.

There's nothing we can do about it; we'll pick it up later. Stuart squeezes into the Bedford's cab, Teddy and I join the young Africans in the back and we move off again.

As we trundle across flat grassland I hear the humming of a thousand bees. As we draw closer to the Mission the humming swells into the beauty of young African voices singing evening Mass in *Swahili*.

Night is falling. We're welcomed by the Fathers and refreshed with glasses of their home brewed Tamarind wine and a light dinner of rice and vegetables. One of the Fathers tells us that every night a Water Buck ravages their vegetable garden. Could we do something about it? Teddy says he will deal with it immediately, goes out to the Bedford and returns with his rifle and a flashlight. He will kill the bothersome Water Buck for them and invites me to come with him.

Once again the strangest paranoid thoughts invade my mind as I follow Teddy into the night, thoughts of Jean's infidelities, thoughts that perhaps she had already charmed this handsome White Hunter. Teddy stops, whispers he can see our quarry. Tells me to stay exactly where I am and not to move nor make a sound. He creeps forward disappearing into the darkness. The confused mind rages on. Was he in love with Jean? Was this the perfect occasion for an accidental shooting? A hunting accident; I'd heard of such things and waited in cold dread for the bullet. Teddy fired and the Water Buck was dead. The mind. O the mind.

58.

In the morning the dead animal is retrieved and we divide it between the Mission and ourselves. Matt ties our half to the side of the Bedford where it will mature with the help of flies and a variety of other flying insects and keep all of us fed for a few days.

Our passage through the bush countryside is periodically interrupted by engine failure. Jack jumps down, opens the bonnet, fiddles with the old Bedford's intestines and gets it going again. It's evening by the time we reach the river. We park several hundred yards from its bank and make camp.

Camp is elementary. A tarpaulin spread alongside the Bedford where we would sleep side by side. There's no need for any kind of shelter, the air is warm, mosquitoes the only discomfort; there's not much can be done about them. Horace and Petrus put together a small fire for cooking and for deterring any curious beasts of the night from coming too close. They carve off a piece of the Water Buck and roast it in the glowing embers of the fire. After the meal and a smoke, sleeping bags are unrolled on the tarpaulin and, exhausted from the drive, we're ready to crash. Horace and Petrus curl up around the fire chatting to each other in native tongue.

59.

The morning starts with coffee. Coffee? We call it coffee. Its colour is the same as the river and, brewed in its rusty brown water, carries with it a rusty, brown taste.

Then there's the need to shit, not too close to camp and armed with a rifle in case a wandering predator catches you with your trousers down. I'm intrigued by the way nature takes care of the excrement and watch fascinated as Dung Beatles roll away pieces of shit as big as themselves to their dens to give continuity to their specie.

Shaving is a sometime thing. Jack never shaves, nor do I. Matt grows very little facial hair. Stuart shaves every other day around his moustache. Teddy shaves his face every day with care.

Breakfast eaten, toilet completed, we prepare to start work. There are only three rifles and Stuart's piece of double-barrelled artillery so I remain an unarmed observer. Jack says perhaps I could take a shot some time but bullets are costly and not to be wasted. Rifles are loaded and led by Jack we quietly approach the riverbank.

The surface of the slow flowing brown river is unbroken. We wait a few minutes to see if any creature would break the surface but there's nothing, no sign of crocodiles. We move on. The bank undulates as it follows the curves and bends of the river, sometimes high above it sometimes level with it. We pause at a high point as a herd of Hippos leave the pastures where they've spent the night grazing and now wade the river beneath us.

Unbelievably, Stuart is cocking and raising his huge rifle to aim at the back of the head of one of these unthreatening animals. It's close range. He fires killing the beast instantly. The rest of the party are surprised but undisturbed by this useless killing. They joke about itchy trigger fingers. Stuart, exhilarated, says the cartridges he was using were old and he had to be sure they still worked. The huge body of the dead Hippo, belly up, floating away downstream, is proof of that. Horace wades into the river with a rope, loops it over one of the Hippo's vertically erect legs and pulls it to the bank.

Captain Stewart Stuart is a sham; killing that huge beast to test his ammunition. What nonsense. Word of the slaughter spreads quickly to nearby villages and within a few minutes fifty or so jubilant villagers are surrounding the dead Hippo hacking it to pieces and carrying off sizeable portions of its flesh. Horace hacks off a few slices for us. Nothing goes to waste.

In the evening Petrus cooks up strips of the Hippo meat. It's tough as leather tasting vaguely of pork. Over a smoke, Teddy regales us with stories of derring-do, actions and adventures he'd experienced in the Malayan conflict when he was a subaltern in the Ghurkhas.

The following day, we creep to the bank of the river and conceal ourselves and wait. We see our first crocodile. Jack hisses there's one close to the opposite bank. I strain my eyes to see it. Where is it? It takes me some time to make out eyes beneath a horny skull and the tips of nostrils just above water. In a whisper Teddy tells me that when the crocodile has taken enough air it will sink to the bottom and stay there for about twenty minutes before coming to the surface to breath again. While he's on the bed of the river we will cross and position ourselves on the far bank to get a better shot at him. I'm not happy at the thought of crossing the river knowing a croc is holding its breath a few yards away. Teddy assures me a crocodile can only take its prey by coming to the surface and we have guns.

The crocodile submerges. We remove all our clothes except for plimsolls to protect our feet from the rocky bed of the river. Stark naked, clothes bundled and held with one hand on top of the head, a rifle gripped in the other, we file into the river. With considerable trepidation I follow and wade chest deep to the opposite bank. What a bizarre sight we are! Jack in his wife's old felt hat and pink plastic rimmed sunglasses, a break in the frame mended with an Elastoplast. We all wear hats of one kind or another and belts holding our knives around naked waists.

Arriving at the other bank, we clamber out of the river. Matt is to be the marksman and he sets up a position in clear view of the river where the crocodile is expected to appear. He cuts a mount for his rifle from a tree branch, sets it firmly in the ground and lays his loaded, cocked rifle across it, safety catch on. Out comes the tobacco and rolling papers. We all sit back and have a smoke while we wait for the crocodile to return to the surface and its nemesis.

60.

The callousness disturbs me, the practiced, organised nature of the operation. This wasn't Man the hunter-gatherer killing for food and survival it was a commercial enterprise and I was there of my own free will for kicks. Teddy told me that when he worked for a Safari Company in Arusha, if a client wanted to kill a Leopard or Cheetah for a trophy, he would first kill a Baboon or Warthog for them, hang the carcass from the lower branch of a tree, then secrete the client with a rifle in a hide close by. He would wait by their side in case anything went wrong but soon, a Leopard would be attracted to the bait and while sniffing around was an easy target.

61.

The crocodile arrives on time. No one moves but Matt. He slowly raises the rifle butt to his shoulder, slips the safety catch, aims carefully at the bony skull of the beast protruding no more than two inches above the water and fires. The crocodile surprised in death thrashes around for a few seconds then, rolling over belly up, starts floating gently downstream. Horace and Petrus hurry into the river to retrieve it and drag it onto the bank. If left to drift, the heat of the sun would soon ruin the belly skin.

I learn how to skin. I want to be part of the safari not just a detached observer. Jack thumbs the blade of my 'boy scout' knife. Shaking his head he strops it on a piece of carborundum he carries with him and returns it to me sharp enough to slice a hair. The dead crocodile, about fifteen feet of it, lies on its back. I cut into the soft skin where it meets the bony armour protecting its back and start lifting the precious skin away from the layer of white belly fat. Jack works from the other side instructing me. As we work, Marabou Storks and Vultures fly in, settling in a wide circle around us. They edge forward impatiently as the skinning progresses. The skin, freed from the body, is carefully rolled up. As we move away, the carnivorous flock rushes to the naked body and tares at it in an eating frenzy. Nothing goes to waste. Petrus and Horace carry the skin back to camp where it's unrolled, liberally salted, rolled back again and stored in the Bedford.

62.

Fear recedes. I'm getting used to living in the same water as Crocodiles. At night they retire under the banks of the river. Wearing head-torches we wade the river seeking them out. Eyes gleam back at us presenting easy targets but there's no way of telling the size of their owners; we try to avoid hitting young crocs too small to skin.

Night hunting has other hazards. Just getting to the river can be dangerous. In the daytime, as we trudge across the bush from one bend in the river to another, long, thin, venomous Black Mambas sensitive to our approach suddenly whip up above the grass and speed away. At night Puff Adders, notoriously slothful, extremely venomous creatures, choose to go to the river to feed and drink and we often find them in our path. To accidentally step on one could be fatal or at least result in a painful wound.

Jack says he's been snake-bitten more times than he can remember and survived but it's something to be avoided.

63.

There are critters that prefer night to day. We see eyes of all kinds reflected in the light of our head-torches. They can more or less be identified by their height above the ground. Rodents and foxes are close to the ground; up high, Nightjars, nocturnal birds, hiding in the trees, blink at us in the light. Eyes that glitter at middle height could be more serious, possibly a Cheetah or Leopard.

We're returning from the river, Jack identifies the eyes of a Leopard that seems interested in us and is following. Jack is leading the party, Teddy and I behind him in line, Stewart is not with us, he's in camp nursing his wobbly leg, Matt is bringing up the rear. There's a rustling, a growling and snarling. Matt swings around and fires from the hip at the Leopard as it breaks cover to attack. Matt's bullet catches the beast full in the throat and it dies instantly; a beautiful female Bush Leopard, tawny spotted fur on her back, soft white belly fur.

Jack says it's strange for her to have attacked us. The following day we're going to return to skin her - a trophy for Jean and me to take back to England. When we skin her we discover a huge festering bruise in her side; she had been gored and made angry enough to attack anything that moved.

Safari is not a comfortable way to pass time. Primitive accommodation, made even less comfortable by every day irritants, tsetse flies by day and mosquitoes by night. I've seen movies of Hemingway stories enacted by glamorous movie stars slumming it in the bush in fancy tented camps, swilling champagne with their caviar served by white-coated, *tarbooshed* African waiters. We don't even have a chair to sit on and certainly no booze or glamorous stars.

64

Sometimes, of an evening after a day's work, I'm given one of the rifles and a couple of the dubious rounds of ammunition, the ones with cotton thread tied around the base, to go and shoot something to eat while the hunters relax in camp. Small game is plentiful and my aim good. When I fire, clouds of smoke burst from the ill-fitting cartridge in the breach obscuring my view. Only when it clears can I see if I've hit anything.

Relaxing in camp, having a smoke after a meal of tough meat, rice and a cup of brown water coffee. There's a pile of magazines and old newspapers either to read or use as cigarette rolling papers or as toilet paper. I leaf through them, surprised to find some of the stories that Teddy had told as his own experiences. Some people are strange that way; making up romantic stories of derring-do even when the life they actually lead is so hazardous.

There are days, moments when I feel close to danger. In constantly crossing the river, to be sure of my footing I had cut for myself a stave to prod the uneven riverbed. Preparing to cross the river I prod what I think is a rock by the bank and it slides away from me.

I find myself on a spit of sand protruding into the river photographing Matt on a high point firing at a crocodile beneath him. He hits it but not fatally and it's angrily making haste in my direction. Matt finally clobbers it but there was a moment.

65.

Today on our way to the river Teddy identifies a large pile of dung as elephant shit. He still has a license to kill one. With the settling of increasing numbers of farms in the elephants' habitat some farmers will kill them in defence of their crops. Licenses issued to kill them are seen as a necessary cull. We spend the rest of the day tracking this one down, one pile of shit after another. After an hour, Teddy finds a pile that is still warm. We're getting close. Torn tree foliage sticky with sap shows the elephant passed there for a feed not more than a half hour ago. The next load of shit is straddled. The elephant has wind of us and is no longer stopping to shit. We follow. We find him and his whole family, a dozen elephants peacefully grazing in a clearing. We pause. Teddy grabs a handful of dirt. We are downwind of them but still approach carefully, he sprinkles the dirt to confirm wind direction as we move slowly towards the herd. The herd takes no notice of us. Teddy whispers there's no big ivory worth the trouble. We retire and the elephants graze on.

66.

After three weeks on the river and two-dozen skins rolled and stored, clouds are appearing in the sky, dark clouds heralding the rains; this maybe our last night. We're camped and bedded down a hundred and fifty paces from the river when it starts to rain and didn't it rain! It poured!

We wake next morning flooded, almost in the river you might say. Everything's soaked. Not waiting to make breakfast, we quickly pack the camp's soggy remnants, pile into the Bedford and start to squirm our way across muddy fields back to Sao Hill. Ditches and streams we'd rolled across on our way in are now raging little torrents to be carefully negotiated. Sometimes we enlist the aid of local tribe folk to help pull us through and we pay them for it. The Bedford coughs, splutters and dies a few times but Jack's administrations keep it going and we arrive back at the motel. Emma and Jean come out to greet six exhausted men.

Emma, in cheerful mood, immediately leads the Mattheisens, Stuart and Teddy to the motel bar; the two Africans slope off to their dwellings at the rear. It's nothing unusual for them, possibly anti climactic now they're back to being sometime motel servants. I give Jean a long hug. For me it's a moment of exultation. I've survived a dangerous macho game; she's glad to see me back in one piece. I'm happy to have her in my arms again. I'm the great white hunter with the skin of a leopard returned to tell the tale. That's how I feel.

67.

The motel gang is dispersing. Jack and Matt have taken charge of the skins and have already left for Nairobi with them in the Bedford to sell them to leather manufacturers. Teddy is headed for Arusha to take up a tourist safari job. Stuart and Emma are left alone to retrieve the jeep and tend the needs of passing commercial travellers; we're back on the road to Iringa headed again for Dar Es Salaam. We'd never been there before but are drawn on by the promise of its name, House of Peace and the pleasing prospect of finding an oasis of Arab culture.

In Iringa we thumb a large truck carrying produce to the coast, to the House of Peace. The driver is a young, affable, talkative Indian. The journey, about five hundred miles, takes almost nine hours and we arrive in the middle of the night. He says he lives alone and it would be a great honour if we would accept his invitation to spend the night at his home. The further we get from western style organised life the more we become aware of people's easy sense of hospitality. We unload. He takes us upstairs and shows us into a plain room empty of furniture and wishes us a good night. This is where we bed down to sleep.

We unroll our sleeping bags on the bare floorboards, zip them together and get in. A few hours later, I'm awake. There's enough light coming from the street for me to see, wall to wall, a heaving ocean of cockroaches completely surrounding our bed. Jean, disturbed from sleep by my movement, looks around her. Unable to deal with this entirely new insect phenomenon she shudders and hides back inside the sleeping bags. I join her. We may have seen a cockroach or two before but we'd never experienced being completely surrounded, overwhelmed by them in their tens of thousands like this; they don't attempt to invade our sleeping bags and in the morning are gone like vampires hiding from the daylight.

68.

We go to the market and buy food for the day; make cheese sandwiches; sit and eat one while we drink a cup of fine strong Turkish coffee and consider our next move. It has to be Zanzibar. We have no tangible reason to visit Zanzibar. We're being drawn there by the same sense of mystery that brought us to Dar Es Salaam. How to get there? Dhow. The Dhow has a thousand year history of plying the coasts of Western India and East Africa, blown in a grand circle by the Monsoon, spreading trade and culture. We go down to the harbour and view our options. We negotiate our passage arguing the price of the fare with the skipper of a large vessel. We set sail in the afternoon on the seventy-mile journey.

Close to the Equator night falls abruptly; we start eating our cheese sandwiches and they begin creeping out. Within minutes the entire boat is running with cockroaches. They're crawling everywhere. We hide the food. Our experience the previous night does nothing to lessen our disgust for these Jurassic creatures that inhabited the planet long before we got here but we don't philosophise or throw ourselves overboard. We maintain, wary and awake, until dawn when they all disappear back from where they came, down cracks between deck planks, back to anywhere on the boat that's deep and dark. We finish the sandwiches for breakfast.

Zanzibar, long before we can see it, breathes out an aroma of cloves that wafts across the sea to us. It's a small island, the port a labyrinth of alleyways, spice markets, narrow streets and old men riding donkeys with their feet dangling at the side. It's a mix of Arab, Indian and African influences, whitewashed stone buildings; brass studded elaborately carved wooden doors. We wander the town weighed down by our rucksacks looking for an hotel and pause to rest at a table on the terrace of a café. We need something cooling to drink and order beer.

We attract the attention of a western-dressed, middle-aged Indian; he invites us to his table and buys us another beer. He's polite and well educated; Harrow and Oxford. He's curious about us and we fall into conversation about our trip. Back in the world of people and custom, our political minds are reactivated; we quiz him about the current state of affairs on the island. He gives us a general understanding of the political situation.

He says there's no overt racism between African, Arab and Indian, but they can't reach agreement on the future status of the island when the Colonial Office withdraws its governance. Amongst the Arabs, minor political dissatisfactions are giving way to a more visceral urge for independence; they want nothing to do with the Colonial Office's democratic Legislative Council and want to see the island become a Sultanate.

He invites us to stay at his beach home on the south of the island. A charming host; the beach, broad and soft with white sand. We stay a while and enjoy his hospitality.

Seashells are a major industry here, sold to be ground down as fertilisers. Before leaving the island we visit the seashell market, a plethora of colour, shape and pattern; we buy a few of the more unusual species and tuck them away, along with the leopard skin, in the bottom of my rucksack.

69.

We make our way back to the mainland the way we arrived. Knowing what to expect lessens the fear and discomfort of the nocturnal cockroach invasion. Once on the mainland we head for Dodoma, about four hundred miles north west of Dar es Salaam. We hear poetry in the names of the towns we pass through and make up a song celebrating them. We're going to Nairobi. *Na kwenda Nairobi, na kwenda Nairobi. Mbeya, Iringa, Dodoma, Arusha, Nairobi...*

It's raining in Dodoma. We remember we're supposed to check in with the District Officer and think it a good idea to do it now. He's angry. Where the hell have we been? We'd let the DO in Mbeya know we were on the road to Iringa and since then there'd been no word from us. Didn't we know we were expected to check in at each destination? They'd sent out search parties looking for us. We'd caused unnecessary work and frustration. He's really pissed off. After a few moments of embarrassing silence he says he doesn't know whether to arrest us or take us home for dinner.

He and his wife, both in their mid thirties, live in a modest European-style house on the edge of town and are perfect hosts. We wash up and change clothes and sit down with them to an English dinner while our discarded, sweat-stained khakis and underwear are cleansed of the past in their washing machine. There were not too many opportunities to freshen everything.

He talks about the imminent decolonisation of the territory. Within a few weeks Tanganyika will become Tanzania, independent of the Crown and Julius Nyerere will be its first African president. The Colonial Office, despite being considered responsible for long time suppression of African rights, is doing everything in its power to ensure transition will be effective and peaceful. He's proud to be the District Officer at this time in history.

70.

He has a problem, one he thinks we might be able to help solve. Two days ago a young Dutchman arrived at his office in a brand new Volkswagen beetle. He'd come from the Congo; for the last two years he'd been a carpenter in a copper mine. He'd accumulated a bit of cash, bought the car and was intent on driving it home to the Netherlands. He was new to driving; hadn't yet passed any kind of test; besides that, the unpaved graded earth roads were swimming in mud after the rains; he hadn't a chance in such a lightweight vehicle and couldn't be allowed to leave on his own. The DO thought perhaps we might help out this unfortunate Dutchman as well as ourselves by being his main driver and providing serious ballast against the road of mud. A lift to the Netherlands? We're keen to help.

Meeting him the following morning at the DO's office. Hank. A tall, well-built man of our age dressed in khaki shorts and shirt wearing a bushranger's hat, broad brim cocked on one side to accommodate a shouldered rifle even though he had no rifle. He speaks very little English and can't make head or tail of the few words of Afrikaans I'd learned in South Africa; even so we manage to communicate enough to understand we can be of benefit to each other. The DO explains to him he can only be allowed to continue his journey north if he takes Jean and I along with him. Hank looks at us, shrugs and agrees. An hour later I'm at the wheel driving the three of us north through the mud to Arusha. The DO is pleased to see us go. He's clobbered two birds with a single stone and we now have a lift to Holland.

71.

It's near four hundred miles from Dodoma to Arusha and we're making fairly good time. The road is slippery but the Beetle with our added weight copes with it. As we near Arusha, Hank says he wants to pause there and climb Kilimanjaro. We don't want to climb Kilimanjaro. We'll make our own way to Nairobi and he can catch up with us in a day or two. Leaving Hank now, we're unsure of ever seeing him again or that easy ride to Holland.

We are destined, in Nairobi, for the office of J Walter Thompson where I can do some work for them and replenish our dwindling funds. In Arusha we unload our baggage. None of us knows Nairobi so we give

Hank the address of the J Walter Thompson office; they will know where he can find us. We watch him drive off jerkily in the direction of the mountain leaving us stood on the road looking for a lift. Within five hours we're in Nairobi.

It's February 1958. White Kenyans are still recovering from the horrors of the Kikuyu Mau Mau uprising that had held them in fear of their lives for more than five years. Some people are still wearing side arms as if they aren't quite sure it's all over.

We find the JWT office and meet the Managing Director, Brian Robertson, a friendly young Englishman. A bachelor, he invites us to stay with him while we are in Nairobi; he'd been informed of our possible arrival by the Johannesburg office. As it happens, he's in need of help and we're running short of money; he's keen to set me to work on some of his accounts.

We follow up an introduction from Sylvester Stein to the East African stringer for Drum Magazine, Alan Rake, a man our age. He's a mine of information about local politics and the contest for power between the Kikuyu and the Luo people. He's become friends with the leader of the Luo cause, Tom Mboya, and introduces us to him. We attend Tom\s inaugural speech at the Legislative Council and the celebration that followed. He's a bright, upstanding young man, Oxford educated, confident of his destiny in Kenyan politics and can dance the Foxtrot.

72.

Arriving for work at the JWT office the following day we see the mud-splattered Volkswagen parked outside and Hank, good as his word, asleep at the wheel. He's pleased to see us. His drive from Arusha had been a nightmare, slipping and sliding in directions he didn't want to go, bumping into other vehicles and several times needing help to tow him back onto the road.

The designs I'd prepared for Brian Robertson are finished and paid for and we're ready to get back on the road. Apart from meeting Alan Rake, Nairobi held little attraction for us. Of course there's no urgency but the road beckons, the rainy season is almost over.

The weather has improved but not our chances of making it up the Nile. Egypt has still not resumed diplomatic relations with Britain; our Britannic Majesty's passport won't work; Hank could get through on his Dutch passport but was hesitant to make the trip alone; we have to revise our route.

I consult the highways and byways bible. It looks like a hard drive northeast across the Northern Frontier District, over twelve hundred miles of scrub and rocky desert. We drink our farewells with Brian, pack our baggage into the car and set out with Hank on a much drier dirt road to Nanyuki, a hundred miles into the desert northeast of Nairobi. I'm driving.

Nanyuki's a small town and last railway link out from Nairobi, its most notable address, the Silverbeck Hotel. It sits on the outskirts of town, fair and square on the equator, this geographical accuracy verified by a brass strip angled across the bar in the saloon. We're stopping here to stock up with enough food, water and gasoline to get us to Moyale on the southeast border of Ethiopia. Our next map coordinate, Isiolo, just fifty miles down the road.

The Isiolo District Office. A white, two storeys fort-like building, surrounded by a dusty disorder of whitewashed dwellings. We report to the District Officer and announce our destination. He seems to be certain we'll make it, so certain he asks could we carry a dispatch for him to the embassy in Addis Ababa. Why not? An official dispatch could hold us in good stead if we encountered any kind of official trouble. I tuck it in my backpack.

73.

My relationship with Jean, real and imagined, has settled into the kind of caring friendship we had sometimes known. Hank holds no fascination for her and daily uncertainties keep us closer together, dependent on each other more than before.

Checking again with our Highways and Byways bible, the trail to our next destination, the Ethiopian border town Moyale, looks uncertain, some parts of it displayed by faint dotted lines indicating 'Hippo path' or 'Dried up river bed' Our plan is to drive as close as we can get to the start point of a metal road to Addis Ababa built by the Italians during their colonising war against Haile Selassie. The nearest point is the battle-worn town of Neghelli captured in the war by General Rodolfo Graziani, known as the *Marquis of Neghelli,* a title bestowed on him by Mussolini for his military success. We set off for Moyale, a distance of about four hundred and fifty miles of flat, rocky scrub desert, home to a rare species of antelope, the long necked Gerenuk; we are fortunate enough to see one.

Driving along a desert trail, visible only by the imprint of traffic, we sometimes come to a faint suggestion in our path of a fork and have to decide which trail to take. We get out of the car and, like hunters tracking an animal, carefully examine the ground for the heaviest imprint, the most used direction, make a decision and carry on. A few miles out from Isiolo we're confronted by the possibility of a fork and, after minute inspection, take the wrong trail. We drive about two hundred miles before we realise.

we're on the road to Marsabit, not Moyale. It's evening; the light rapidly failing. Frustrated by the amount of time and petrol we'd wasted, we turn around and head back to Isiolo.

Driving at night and already exhausted, I fail to avoid a rock that shears the Beetle's sump plug, unnoticed until the oil-less engine grinds to a halt. We're stranded in the middle of the Northern Frontier District desert in the middle of the night unable to move and no one to shout to for help. It's not a situation any of us had expected. We don't know what to do. We lock ourselves in the car and slump into fitful sleep.

74.

It's morning. We breakfast on the water and food we have with us. There's nothing we can do but sit and wait. Wait? Wait for what? Mid morning we see a cloud of dust on the horizon and the sound of an approaching vehicle; a Landrover. A unit of the British Army; six soldiers and a subaltern, a patrol that keeps a check on border infringements by neighbouring Turkanas and Somalis. Their young officer inspects our credentials then takes a look at the Beetle's engine. He doesn't say a word, looks at his watch and returns to his troop. They remain seated in their Landrover. The officer looks at his watch again but says nothing. We wonder what the hell's going on. All we can do is to sit patiently in the Beetle.

The distant sound of another approaching vehicle gets the troops out on the road. This time it's a large, heavy duty Italian truck. The young officer stands in its path hands raised; the truck stops. It's empty and headed for Nairobi. The officer has a few words with the driver who then manoeuvres the truck, backing it up to the Volkswagen. The soldiers release its back flap. Three of them on each side, they lift the Beetle and slide it onto the bed of the truck. The three of us get in the back with the car; wave goodbye to grinning soldiers as the truck pulls away.

The driver takes us back to Nanyuki where we unload the car onto a railroad platform and roll it down a ramp onto the road. We find a garage mechanic; he's on the telephone ordering a new engine from Nairobi while we repair to the Silverbeck Hotel's bar on the equator. Downcast at the fix we're in, we contemplate our future over stiff drinks. We wait. We wait for ten days. My rucksack is beginning to stink. I think it must be the leopard skin stuffed at the bottom. The new engine arrives from the Volkswagen dealer in Nairobi. Our local mechanic fits it and we get back on the road, back through Isiolo, on to Moyale.

75.

As we get closer to the Ethiopian border the desert takes on a different character, large rocks abound and have to be negotiated. The rougher road requires low gears and consequently a higher consumption of petrol.

Moyale is home to a few hundred villagers and an Indian *duka* attended by a genial young man set down there by his Nairobi family to learn the business. It's a lonely life but he doesn't complain. We go shopping amongst his shelves of canned foods. We buy cans of beans, sausages and soups. He offers us shelter for the night at no extra cost but we decline. The nights are warm and we enjoy camping out in the open. We need to build a fire to cook our supper. We find a pleasant meadow where we can comfortably camp

I, with my experience of safari, take charge. I have overweening confidence. I've heard what to do when making a fire. I've seen it in a film somewhere. The grass is long and dry. We have to burn out a patch so the fire won't spread out of control. Jean is startled. What kind of fool am I? She shrieks at me. Burning out a patch? You're crazy! The whole place will go up! Intolerant of her criticism of my field craft I insist she goes back to the car. I set about clearing a couple of square meters; I strike a match and drop it into the grass. There's a dull thud like an explosion as if I'd dropped the lighted match into petrol and within seconds the fire is out of control and consuming a huge area. Locals rush out as if from nowhere armed with tree branches and start to beat it out and they soon have it under control and extinguished. They don't say a word. No recriminations, they just depart. They must have done this before.

Shame is hard to bear. There's nothing to say, the smoking remains of an entire meadow says it all. We eat the beans and sausages cold, straight from their cans and bed down for the night. None of us speaks. We're up at sunrise, packed and back on the road. The inhabitants of Moyale must have been glad to see us go; we drive into Ethiopia headed for Neghelli, two hundred and fifty miles further north. The road is extraordinarily rocky, worse than we could have imagined. At times we are driving along dried up riverbeds, those paths indicated on our maps as faint dotted lines, the Volkswagen without fixed chassis, torques its way snake-like across large boulders. It's a hell of a drive in the lowest gear and consumes a large quantity of petrol, more than we'd allowed for.

76.

The petrol gives out. We're less than twenty miles from Neghelli, night is falling. There's only one option; one of us is going to have to walk the rest of the way to Neghelli and bring back help. Hank is loath to leave his car; Jean is not keen to wait with him so it falls to she and I to make the trek. Hank hands me a tiny Berretta .22 automatic pistol he's had secreted away and a matchbox full of bullets for it. About to trek eighteen miles at night in lion country, this puny firearm doesn't do much for my feeling of uncertainty about the task in hand. We split the water supply between us, I take the dispatches we're carrying, tuck them along with our passports into my jacket pocket. Jean and I set off down the road leaving a very apprehensive Hank wondering when or whether he'll ever see us again.

Having read about the Italian defeat of Ethiopia and the fall of Neghelli to the Italian army under Graziani, we expect the town to have been Italianized. We're looking forward to relaxing in an Italian-style café, eating a *pasta carbonara* and swilling a fine bottle of *Chianti*. The road is hilly and as we top each hill we half expect to see the gay lights of the town and hear the strains of café music. We top hill after hill, the horizon remains dark and blank; the only sound a distant grunting of lions.

It's about four in the morning and still dark when we suddenly find ourselves in what must be the town itself. No lights; no café; no music. Certainly no *spaghetti carbonara* or *Chianti*. Dark, stockaded houses line the road; the only sign of life a dog. He barks as we pass. We don't know where to go until we come across whitewashed rocks that indicate a turning off the road. If someone took the trouble to paint the rocks white they must lead somewhere. We follow them to a high and heavy wrought iron gate. No building, just the gate.

Suddenly from behind the gate a voice, an aggressive command in a language we don't understand and the ratcheting of a rifle bolt stops us in our tracks. Out of the darkness barefooted figures in army greatcoats emerge carrying old long-barrelled Lee Enfield .303 rifles and surround us. They are not friendly. They shout at us and indicate we should sit down on the ground. I drag our passports and the dispatch from my pocket. They are taken from me, examined and obviously not understood. They exchange comments among themselves. We sit apprehensive about what might happen next. Nothing happens. We wait. If I'd had any army experience I'd have realized that only the military would whitewash rocks and that we had stumbled into an army encampment. We wait.

77.

The sound of an engine; a jeep appears driven by a young army officer. He gets out, has a word with the soldier that has our passports and the dispatch, takes them from him and walks over and returns them to us. He addresses us politely in Italian. In the course of her university education Jean had learned some Italian. She responds to the officer's questions explaining our situation. He invites us into his jeep and we drive off. He takes us to a Norwegian Lutheran Mission. It's the only one of its kind in the whole of Ethiopia and happens to be in Neghelli.

Dawn is breaking. The officer knocks at the door. After a few moments, the missionary, a white haired gentleman of about sixty, opens the door rubbing sleep from his eyes and invites us in. The officer greets him, says a few words in *Amharic*, salutes and leaves. The Missionary listens sympathetically to our story. Seeing we are absolutely knackered he insists we get a few hours sleep before taking us back to the Volkswagen with a can of petrol. He gives us blankets and finds us a space to lie down. As I fall asleep I'm thinking of Hank sitting locked in the car, waiting, wondering.

We breakfast on coffee and bread. Our host isn't in a hurry; doesn't seem to share our anxiety about Hank, waiting for us, probably thirsty by now. He's taking his time; talking incessantly. He shows us his little museum of local tribes' arts and crafts. In one corner, dangling from a rack, numbers of small clumps of withered and blackened flesh. I ask what they are or were. He replies with academic detachment. They're the remains of genitals of invading Italian soldiers taken by Galla tribesmen during Mussolini's invasion of Ethiopia. He fills a two-gallon can with petrol; we refill our water bottle with fresh water, get into his Land Rover and start the drive back to Hank.

As we approach, Hank bursts from the car almost delirious, waving his arms excitedly. Shaking his head and sighing deeply a number of times, he babbles his relief in Dutch. He's very pleased to see us. The missionary empties the can of petrol into the Volkswagen's tank. He says we should stop by the Missionary on our way through to top up. Blessing our future, he gets back into the Land Rover and drives off.

We follow him in the Volkswagen back to the Mission, top up the tank and head for the metal road, the heaviest black line in our highways bible. It's two hundred and fifty mile to Addis Ababa, an easy five-hour drive on tarmac.

78.

Addis Ababa? We have no idea what to expect. We enter the city limits and find there is no city, only a widespread fragile network of dwellings and shops but with no obvious central features that might be expected in a Capital city. Were driving down an avenue called Winston Churchill. Numbers of elaborate wrought iron gates line the avenue suggesting an affluent neighbourhood but there are no grand houses behind these gates, only cows grazing peacefully. It's trying to be a city but something's holding up its development. We find the Emperor's palace, the only building of note, guarded by caged lions. The only truly modern construction of concrete, steel and glass is the commercial bank; a veritable shining temple of Mammon. Apart from these there seems to be little else of architectural interest. It's early days for Addis Ababa.

We find our way to the British Embassy and are interviewed by a member of the consular staff who wrinkles his nose at the now disgusting smell emanating from my rucksack. We learn there has been no change in the diplomatic relations between Great Britain and Egypt. Our journey is at an end; our only way home is to fly. Hank would have to find his own way. Being Dutch, the Sudanese and Egyptians had no quarrel with him; he was free to cross their frontiers. He would try to keep to our original plan; find the Nile, follow it to Alexandria and then grab a boat to mainland Europe.

We'd shared more than two thousand miles of adventure with him. We'd grown accustomed to being in each other's company. We'll remember each other.

We meet the Ambassador and hand over the dispatches we'd been carrying from the District Officer in Isiola. He thanks us for them and asks if we play bridge. We don't. Disappointed, he sighs, shakes his head and hands us into the care of Tamrat a young Ethiopian member of the Embassy staff. He takes an interest in us. We become friends; he invites us to stay with him.

Addis Ababa holds little charm for us. We've arrived. We'd done it. Our journey is over. We no longer have enough money for two BOAC tickets and have to surrender our passports and prepare for repatriation.

We give Peter as our reference. The Embassy telephones him and he agrees as our lawyer to guarantee repayment of our fares. I'm keen to get back to London despite the possibility hanging over me of being conscripted into the army. I don't give it much thought; it's just something to be avoided.

We have a few days to kill while embassy bureaucracy records our existence and books our flights. We wander street markets with Tamrat looking for gifts to take home. Thinking of Peter, we find a primitive painting of the meeting of King Solomon and the Queen of Sheba the story told in a series of frames like a graphic novel.

Tamrat, his name means 'miracle', explains Ethiopia's interesting history of religious beliefs. There's little doubt the meeting of Solomon and Sheba and their son Menilek spilled Judaism into the mix. Tamrat wasn't sure of the exact date but he thought St Frumentius brought Christianity to the country sometime around the third century. In the sixth century, the Empress Judith of Ethiopia converted the entire population to Judaism lasting for the duration of her reign. All that remains of Judaism today is a single tribe known as *Falashas* who still retain Jewish customs and beliefs.

79.

Ironically, our flight touches down at Cairo airport to let off some passengers. We were not allowed off the plane; it leaves almost immediately for Heathrow. For the rest of the flight we are more or less silent, occasionally glancing at each other, grinning at thoughts of what we had done, where we had been.

Arriving back on home turf arouses in us a feeling of exhilaration and achievement. The possibility of my conscription into the army is still smouldering at the back of the mind; I still don't pay it much attention. We'd completed a journey that will become legend and amaze future children and grandchildren.

Ever loyal Peter is there to meet us. He's as pleased to see us as we are to see him and our mutual pleasure bubbles into laughter of relief. He drives us in his green Morris Minor back to the flat we'd shared near Streatham Common. We chatter excitedly at first, then fall into comfortable silence still barely believing we've made it home. Peter comments on the smell.

In the flat we unload our rucksacks. The leopard skin was innocent; it was the seashells that had been stinking ever since we'd bought them in Zanzibar. I immerse them in water with disinfectant to restore their odourless beauty. Peter's delighted with the *Amharic* version of the meeting of Solomon and Sheba.

80.

Sylvester and Jenny Stein's arrival in London preceded our return and by the time we get back they have already bought a large house on Regents Park Road opposite Primrose Hill. Faber and Faber is publishing Sylvester's book, *Second Class Taxi*, and we're just in time for the celebration. Jean and I bring Peter along to a publishing party that would change his life.

It was an extraordinary gathering, a mix of self exiled Black and White South African dissidents whom we already knew; a number of Sylvester's relatives already living in London; much of Jenny's complex family, her sage and myopic dad, Alan Hutt and her seventeen year-old step brother Sam Hutt hugging a guitar, recently returned from a school trip to Ceausescu's Romania. There were people of fame, Sarah Vaughan was there, a few writers Sylvester had come to know through association with his publisher and the warm and modest John Dankworth and Cleo Laine who had become locked into the anti apartheid spirit. It's a lively, exciting gathering; many new friendships forged. One of Sylvester's relatives, a sexy, eager young woman, Beverly, daughter of one of Sylvester's cousins, is taking a predatory interest in Peter.

81.

Being back in London is a strange experience. Born here and living here I've taken so much of the character of this city for granted and not really looked at it. As an artist I'd made detailed drawings of some of the elaborate Victoriana that graced the city, but growing up in a place, you can become so accustomed to its architectural charm that you fail to give it a second look.

So much seemed new to me. I'd grown up south of the Thames, where nothing much happened. The glittering north was young, decadent, stylish and fun. Living with Peter I was still in South London but attending St Martin's took me across the Thames daily where everything was happening. On our return to London we decide to look north of the river for somewhere to live. We're attracted to the centre. We settle for a damp basement flat in Oakley Street, Chelsea one of London's most pleasant areas. We take walks at night along the embankment. The south bank, opposite Chelsea, still has a few old commercial wharves and warehouses. Above one, blazoned large and red against the night sky, the beer logo *COURAGE* shines out. I was certain itt was speaking to me.

Some of the areas bombed during the war have been and still are being reconstructed and the skyline has changed. The flow of London's traffic has thickened. Trams are out, trolley buses in. There are more small cars on the road, bubble cars and scooters, in contrast to the preponderance of large American V8s in Johannesburg.

We've found a place to live and now we have to get jobs to pay the rent. It's not a problem. My portfolio is admired and attracts three offers, one of them JWT. Having worked for them in Johannesburg and Nairobi I give in to a misplaced sense of loyalty and join them. A group Art Director of another agency offering me a job, Arthur Wilson, was disappointed. He'd taken a liking to me. Despite me opting to work around the corner for JWT, our friendship continues; we drink a beer together after work. Arthur attracts me in a number of ways. He's a rabid socialist and his manner and tone of voice show a certain indifference to authority.

I've always tried to make my advertisements different, even memorable. There's a knack to it; success requires a sense of humour, a break with tradition. This isn't what JWT want of me, neither my knack nor my sense of humour. I was to do what I was told. The art directors are expected to produce stunning, persuasive advertisements based on the mundane findings and logic of the Marketing Department who understand nothing about selling the dream. It's so obvious; I feel the need to tell them.

Post war energy is revitalising the Advertising business. There's a tide of novel thinking that JWT knows nothing about, sweeping away old traditions. Vance Packard had changed the whole attitude to creativity in advertising in the publication of his book *The Hidden Persuaders*, but JWT hadn't read it yet. Inspired by Arthur's fearless attitude to management, I thought it would be useful to have a chat with JWT's Art

Director in chief, the man at the top. Talk with him about a more creative approach. I book an interview.

His secretary allows me into his room. He is elderly. His room is dark. Dark wallpaper, dark polished furniture on dark carpets. We'd not met each other before. He's seated behind a large desk but doesn't invite me to sit. He doesn't say a word, doesn't ask me who I am or what I want, just lets me stand there while he looks at me with pale, watery eyes behind rimless spectacles.

This is new to me. I'm expecting conversation but there's no room for it here. I feel humiliated standing dumbly in front of this unfriendly monster. I don't know how to deal with him. Nothing comes to mind. I have to get out of there.

That evening I'm drinking a beer with Arthur. He laughs when I tell him. He wants me to drop JWT and join him as his deputy. He heads a large group handling many major accounts. Among them the British Motor Corporation, Yardley, Army Recruiting… Army recruiting? It hasn't taken long for fate to catch up with me. People these days were patriotically sensitive about those who hadn't bowed to the law of conscription. Into the second pint, I agree to leave JWT to come and work with him.

Jean and I reconnect with art school friends. As soon as we'd found somewhere to live our friendships continued where we'd left off. We're all four years older and no longer students, each of us now makes his or her own way in the world. John Berry, Best Man at our spectacular wedding, is now a potter serving apprenticeship at a well-known London pottery and married to June, a gifted painter from the Slade School of Art; May Routh, a fashion illustrator married to Adrian Bailey, himself a sensitive painter and illustrator; Len Deighton whose amazing energy knows no bounds, a veritable Ben Shahn of an illustrator, crazy about cooking and is writing a cooking strip for the Observer as well as a novel; Shirley Thompson, his girlfriend, a brilliant illustrator whose work is as immaculate as Len's is wild. All in all a bunch of friendly, talented people looking to gain a living. Jean is the only one of us who hasn't yet found her place. The group suggests she might like to be their agent, represent their work to Advertising Agencies and Publishing Houses. Jean hasn't yet decided what she wants to do. Was there some great life changing ambition that would take her away from where she is now? She didn't say. She's shy of responsibility, unsure she is capable of being their agent, unsure she'd enjoy conducting a commercial enterprise but is prepared to give it a try. They name the group "Vantage". Jean puts together an impressive portfolio of their work and takes it on the road.

82.

Surprising herself, Jean manages the situation very well. Most of the Art Directors, her clients, are men and Jean knows well how gain their attention. The business prospers and she takes on young Jo Lloyd, drug-damaged son of folk singer Bert Lloyd to help her out – so she says – but it's young Jo who needs the help; she's sorry for him, takes him under her wing and to her bed.

I have no special feelings for him; I find his drug addiction pathetic. Whenever we are in the same room his sad, guilty eyes avoid me; occasionally he glances at me with a sneer of superiority that tells me he's probably fucking my wife. Neither he nor Jean admits to it. Now both of us have jobs and keeping different hours our marital friendship is beginning to unravel again. These are confusing times for both of us, re-entering what was a familiar world at a later time. Things have happened while we were away; life is familiar but slightly out of focus. The unquestioning fidelity that characterised our relationship during our hitchhike trip is evaporating. We no longer see ourselves as a couple but as two people each immersed in their own separate lives, neither of us with any sense of what the outcome might be.

We know Anthony Sampson because of his history with Drum Magazine; a university chum of Jim Bailey, he edited Drum for a year before Sylvester took over. He connects with us. He has just published *Anatomy Of Britain*. A Socialist, he's a leading figure in the Observer newspaper's political direction and curious about the value of advertising in politics. He asks if I would be interested in being an advertising consultant to the Labour Party.

Ironically, CPV, where I now work with Arthur Wilson, is the official, paid up advertising consultant for the Conservative Party and their work is a partition away from my own office.

Sampson also recruits a copywriter, Peter Burns, from another agency, a pleasant fellow with Socialist tendencies of about my age, and arranges for us to meet with Hugh Gaitskill and his shadow cabinet at the House of Commons. A number of Shadow Ministers are there, none of them keen to pursue the use of capitalist advertising techniques to further their socialist aims. We finally encourage them to try it despite the discomfort it causes them. It cost them very little for the single appearance of a third of a newspaper page and that's where it ended. They did not understand the possible efficacy of advertising in the modern world and, even though the writer and I were volunteers, regretted wasting party funds on an expensive third of a page of newsprint.

There are regular parties at the Steins and, as usual, we are asked to sing. They are good parties, plenty to drink, an inexhaustible supply of Jenny's stuffed eggs and more and more people. We meet Mike and Evie Fussell, old friends of the Steins and fine painters. Jenny is so keen on them, the large front room of their house has become a gallery of their work and she's not above pushing for a sale if a guest shows an interest in one or other of them.

There's upheaval at work. My group boss. Arthur Wilson has been sent against his will to take charge of the CPV office in Teheran. Arthur responds to this quasi promotion with a telegram from Teheran to the Managing Director.

"Send information how to treat wounds in back."

That's how Arthur saw it. Betrayal. Not uncommon in the advertising industry. Whatever, I'm now head of his group. His departure elevates me to the front line and I begin dealing directly with the Advertising Managers of our client companies. I work on the British Motor Corporation account, Austin saloons and the 3,000 cc Healey, some Yardley cosmetics and, of course, the Army. Fate had drawn me close to my problem at the fulcrum of my career.

John Profumo, the Secretary of State for War, announces his intention to end conscription as soon as he's built a sufficiently large army of volunteers and it's largely my responsibility to make it happen, to make enlistment attractive and, in doing so, free myself from my own nagging anxiety. I'm dragged off to meetings at the War Office where a General makes clear the numbers of recruits he needs. I'm twenty-eight years old and consider myself a malingerer, on my guard against some who would, if they knew, despise me and see me in jail. But I'm thought to be South African, a satisfactory explanation in these circles for not ever having served in the army.

83.

Because of Arthur's departure and my new responsibilities, I get a raise in pay; we buy a small car and move into a more spacious flat in Queens Gate Terrace, close to Hyde Park. We get a telephone call from Bloke Modisane. He's just arrived in London. He comes and stays with us. He tells us what's been going on. The African musical *King Kong* planned two years ago in Johannesburg is complete and has had it's first run in South Africa, touring the country for two years and playing to record-breaking multi-racial audiences. It's booked for a London production in 1961. Everyone will be here. Nathan Mdledle, leader of the Manhattan brothers, playing the lead role. Miriam Makeba plays a shebeen queen. Hugh Masekela, Abdullah Ibrahim, Kippie Moketsi, Tandi Klaasen, all of them will be here. Todd and Esme Matshikiza and their children are already here. Sylvester's found them a flat not far from his house on Regents Park Road.

It was a delight to see Bloke again and under such different circumstances. He was bent on a theatrical life; fancied himself an actor rather than a politicized journalist. He stays with us for a few days then moves to a friend's flat in Soho to be nearer the centre of things.

At work, the army is looming large. I'm introduced to the Parachute Brigade, the face of the army's public relations. I meet with some of the officers, mostly men of my age. Over drinks in the officers' mess I regale them with tales of crocodile hunting. The question of any army experience I might have had never arises. I get on well with them and they seem to take to me. Whenever they are planning a routine exercise they telephone me and ask would I like to come with them. This budding relationship meets with the approval of my Managing Director.

I've always been a photographer, I can remember in my infancy taking holiday snaps on the summer beach with a Brownie Box camera. Now I have a Rolleiflex and I take it with me whenever I travel with the Brigade. It was not my purpose to take photographs for use in any campaign I might invent but, of course, I photographed everything around me.

The first exercise I'm invited to join is in the deserts of Libya. I fly with them photographing them jumping out of the plane. I'm truly amazed. In order to land in close proximity with their unit they exit the plane in such a hurry the last of them is running as fast as he can to get through the door. I photograph them from the dropping zone, dodging them as they hit the ground around me. Tactical airdrops are from as low as five hundred feet, just high enough to allow the parachute to open and break the soldier's fall. Getting the hang of things, I photograph one or two set-ups that might well be used, Pictures depicting determined young warriors fighting for Queen and country. I plan to use them as huge forty-eight sheet roadside posters, like dramatic movie posters.

Jean has no objection to these trips that take me away from home for a week at a time. She takes no interest in them. She's doubtless glad to see the back of me. She has her own social interests and her business seems to be going well.

I'm becoming increasingly confident with the camera. I'm taking and using photographs of my friends in presentations to my clients. I borrow an Austin Saloon from BMC for a weekend and drive with Len and Shirley out into the countryside. They pose for me around the car for pictures that will become part of the next Austin advertising campaign.

I'm called away by the Parachute Brigade to photograph winter warfare skiing in north Germany; then battling and beating the mud with tank transporters outside Oporto in Portugal; then their parachuting into the Troodos Mountains of Cyprus.

My fear of discovery as a malingerer from conscription fades as the campaign develops, succeeding in attracting thousands of young men to join up voluntarily. My relief is considerable when War Minister Profumo says conscription will no longer be necessary. I didn't plan it. I couldn't have seen the way things were going to work out. Who can? But all of the decisions I made, whatever they were about, resolved themselves into a solution to my problem with conscription. Rather than seeing myself as a malingerer, I now feel I've been of service to the Country and deserving of a medal.

I concentrate on my other clients; my most successful exercises being the launch of a completely new car from BMC, a revolutionary design by Isigonis, the Austin Seven mini. I think the best and most interesting way of showing how revolutionary a design it is, is to make a huge exploded drawing of the car, a traditional way of showing the intricacies of a clever mechanical device. It's the best I've done; a precise visual communication of information at a glance. It will be noticed in tomorrow's newspapers.

84.

I get fired. Half the agency, including the Managing Director, has run off and formed another agency and a new administration has taken over at CPV. New brooms sweep clean and I'm out on my ear with three months severance pay.

At this time Sylvester is trying to get some enjoyment out of his job at Odham's Publishing. Since Editing Drum Magazine in Johannesburg, he's still crazy for the way photo-journalistic magazines communicate a story. He wants Odham's to produce one but the boss turns him down. They're too expensive; not the editorial but the paper and the printing.

He convinces management he can produce readable magazines a quarter the normal magazine size *ipso facto* at quarter the cost. They give him a budget and office space to experiment. He calls me. Asks if I'd like to be a part of the project. I have three months.

We hire a lad to help me. I design, he pastes. In the three months we complete two dummy magazines, one a photo-journalistic magazine we name SMITH, the other a popular science magazine we name LOG. We start major stories on the magazine covers to create reader interest at the point of sale on the newsstands. The day we hand the dummies to the Odham's board of directors for their approval, they're busy selling their business to mammoth publisher Edward Hulton. Not much chance of their survival there and time for me to go back to work.

I apply for a job at SH Benson, one of the larger old-fashioned agencies. The Art Director wants to hire me; we discuss salary. I tell him what I want and he's aghast. Says he doesn't earn that much himself. I get what I ask for. These days my confidence is riding high and I'm determined to make an impression on this company that time has left behind.

Now on a handsome salary I buy a new black MGB. It's the first time I've owned or even driven a sports car. I'm in love with it. I take every opportunity to use it. It's more than a utility; my speedy, low-slung chariot is a fun drive.

Sylvester asks me to join him on another publishing venture. Inspired by the success of the American *Kiplinger Newsletter*, a privately circulated newsletter with insignificant production costs and high subscription rates purveying high quality, financially advantageous information. Right now, here in London, buying and selling of property is of major public interest. With *bona fide* research we investigate those areas in London where there is activity and investment. Property is beginning to sell again. Streets of noble houses fallen into disrepair during the war years are being refurbished and put on the market. Prices and mortgages are in fairly easy reach of anyone with a job. It works. *The Stonehart London Property Letter* becomes essential information for anyone buying a house. Circulation increases.

In one of these areas I discover a house I would like to buy on Regents Park Road a hundred yards down the road from Sylvester's. It's in poor condition but close to the Regent Canal and the green expanse of Primrose Hill. I buy it. Jean and I have a house.

We renovate it and it becomes host to many people. The Primrose Hill area is taking on a certain demographic, attracting people from publishing, media, film, and music. We have a piano in the house, parked there by an old university friend of Jean on his way to some teaching post in India. The piano attracts a pianist, the pianist attracts jazz musicians; there's music in the house. Jean responds positively. She sings.

We're at one of the Steins parties, the usual mix of unusual people. It's not exactly a *party*; it's the opening of an exhibition of paintings, mainly work of the Fussells, hanging on all the walls from the entrance hall to the walls of their large living room. After a quick look around at the paintings, the event inevitably sinks cosily into a party. Peter and Wendy Cook are there and hear Jean and I deliver one of our performances. Peter's taken with her voice. With his friend Nick Luard, he's opening a satirical nightclub in Soho to be called *The Establishment*, and he's looking for talent. He hires Jean to sing the bizarre, socially bitter lyrics written by Christopher Logue to the attractive music of Stanley Myers.

Jean's voice also attracts the attention of composers Carl Davis and Richard Rodney Bennett. Richard thinks of her as a Jazz singer. I think Carl has something more operatic in mind, Brecht and Weill, perhaps.

85.

I'm proud that serious notice is being taken of her talent but it's happening in a world other than mine. I lead a corporate, creative life. I wear a suit to work. My social skills are of that world and I find difficulty in relating to the numerous, idiosyncratic poets, actors, musicians and composers who are now constantly in my home. We live entirely different kinds of lives. Theirs is careless, bohemian, with its own clock. Their day ends around four in the morning; mine begins around seven. The distance between Jean and I is palpable.

Two or three of her friends are examining the contents of a matchbox. What is it? Marijuana. I freak. Marijuana in the house of a responsible advertising executive? Have you gone completely insane? Get it out of here immediately. Her friends look at each other, surprised. I know nothing about drugs. I suppose my anger is really about my separation from Jean's life, even in my own house.

It falls to Stanley Myers to make me see it differently. It's a sunny Saturday afternoon when he calls. Of all Jean's music associates he is the most friendly towards me. He asks what I'm up to. Would I like to hang out? I've always enjoyed the company of gentle Stanley and admire his work. I drive over.

He lives in Notting Hill; his flat is a small, comfortable nest with piano, divan and a low, glass-topped coffee table covered with books, music scores, newspapers, magazines, overflowing ash trays. We chat about things that matter to us. Personal things. I'm impressed by his fearless honesty when talking about sex. It's more than I can do. He rolls a joint, lights it, draws on it and hands it to me. I'm a cigarette smoker so I know what to do with it and nobody's watching. I take a couple of drags then a couple more. I hand it back to Stanley and relax against cushions, thoughts of my present life floating around my mind. Thoughts of Jean and the agony she's caused me these last seven years suddenly seem to have an amusing side. I had been victim of my own insistence on the way things should be. How foolish. I can't stop smiling.

It's a different kind of marriage from the one I had expected. I believe I still love her; sometimes I do, other times I don't, but want our marriage to survive. I spend less time concerned about her possible sexual relationships with other men now that I realize where the responsibility really lay for all those years of angst.

Spending less time in her world, I create my own. My advertising work leads me into a life of drinking and partying with the photographers I commission to carry out my designs; Terence Donovan and Brian Duffy, both men of supreme photographic skill and sexual confidence. They have the abrasiveness of East Enders and the awareness and knowledge of Cognoscenti. I enjoy their outrageous company. There's no domestic schedule to meet at the house so I spend most of my evenings with them. We party in fashionable Knightsbridge restaurants and Soho dives.

I'm meeting for a drink with Donovan, Duffy and David Bailey to celebrate a visit by their friend, New York photographer Jerry Schatzburg. Stories ensue. Bailey, enjoying Vogue success, has just bought the latest, much admired, Jaguar XKE, a simple design of aerodynamic beauty on wheels. David recounts an event when driving the XKE into a parking lot.

He's about nineteen, careless with his appearance, scruffy hair and unshaved. The attendant, admiring the XKE says to him, "Bet you wish you owned it"

86.

Freeing SH Benson from the past is proving difficult. The copywriters, most of them frustrated poets or novelists, refer to themselves as the Literary Department. Selling does not run in their veins. Nevertheless they've evolved a set of rules governing the way an advertisement should be created and they published their wisdom as a pamphlet. When Bob Pethick joins the agency, management puts him with me as a creative team warning him they're probably going to fire me. The best way I can show him what I'm up against is to show him the Literary Department's pamphlet. He flicks through it; takes a cigarette lighter from his pocket and sets fire to the pretentious document. When it came to presentation of ideas that differed from the kind favoured by the Literary Department, Bob, himself a writer, adds his contemporary voice to mine. We begin to win the confidence of management. I'm not being fired.

Unless there's a subject under discussion, Bob's a man of few words. He's unassuming. I find him good company and there's little doubt he feels the same way about me. Our success as partners is largely due to shared values, a shared sense of humour. He alludes blandly to the confused and libidinous life I'm leading. I accuse him of envy.

Through our efforts, SH Benson receive their first creative award. It's good for the agency. It's good for Bob and I. Within the year we're headhunted by a smaller, medium sized agency, Graham and Gillies, to become the joint creative directors on the board with options to buy stock in the company. Farewell SH Benson.

Although thick as thieves at work our relationship never extended to our personal family lives. I believe his wife disapproved of my louche behaviour and never invited me to her home. When Jean and I invite them to dinner Bob comes alone.

When he and I took over the creative department we kept all the staff. Time came when we needed another junior in the art department and there were interviews. We settled on a youngster who'd recently graduated from the Royal College. It isn't his birth name but he calls himself Billy Apple. His work is bizarre and original, his manner self-confident. There's nothing in his portfolio that shows the slightest interest in adverting or any of its many manifestations. He's keen to turn his talent in that direction and confidently assures us he has many ideas; we hire him. He's ahead of his time and his ideas are often turned down. We have a product requiring a brand logo. Billy comes up with a tick - √ - yes, a tick. It seemed foolish at the time.

Jean has a problem. She starts singing regularly at the Establishment Club but her performance is shaky; in front of an audience, nervousness distorts the quality of her voice. She's also been commissioned to understudy Lotti Lenya singing a Brecht opera at the Royal Court Theatre. One evening Ms Lenya has a cold and sore throat; Jean has to take over. It isn't easy for her. Had she been singing at home amongst friends there would have been no problem; she loves Brecht and Weill. But suddenly called upon to sing their work in front of a theatre full of strangers was difficult for her; nervousness got in the way. She stumbled through the songs, losing confidence and certainty of pitch.

I watch her show at the *Establishment* a few times but now, when I'm not gallivanting with advertising friends I spend evenings at home.

David Frost is a frequent visitor. He's desperate to be part of the satirical Establishment team but, for reasons I don't fully understand, finds himself constantly excluded. He's equally as talented and, I suspect because of that, subject to the competitiveness of his successful friends. I spend several evenings alone with him at the house drinking late into the night bemoaning his exclusion while his detractors are enjoying themselves on stage.

Richard Rodney Bennet maintains his interest in Jean's career. He sees her as a singer of Jazz and folksong rather than the complex, offbeat songs written by Christopher Logue and Stanley Meyers. He arranges a recording, *Jean Hart Sings*. Her nervousness disappears in the recording studio and the voice, so admired, returns. The album is good. It's not a smash hit but I think it gives her the reassurance, the confidence to carry on.

87.

Insights given me by a few puffs of Marijuana fade; my feelings about Jean's disloyalty and uncertainty return. We lead increasingly separate lives. We live in the same house and share a bed but see little of each other. There's plenty for the jealous mind to work on. I'm paranoid again. I start looking for clues to substantiate suspicions.

I search our wardrobe; feel in every pocket of every jacket and overcoat. I don't know what I'm looking for and find an envelope in one of Jean's pockets. It contains a letter couched in silly lovers language interspersed with insulting references to me, my character, my personality. I'd had enough. Later in the day I confront her with the letter and tell her to get out. She's not happy. She packs a bag and leaves almost immediately. I don't know where she's gone; I don't care.

I'm living alone now in the gloom of failure and insult. Nothing much seems to matter. I have lovers but the nervous, ongoing drama between Jean and I render me impotent. It doesn't last for long. There seems to be nothing more challenging and ultimately rewarding for a woman than to restore life to an impotent member.

Throughout the marriage I'd been haunted by old-fashioned notions of loyalty and good faith even when I was being disloyal myself. Now it's all over, my hurt has receded leaving an easy sense of freedom as it did when Jean left me in Johannesburg, My neighbour and friend, artist Dante Leonelli and his girlfriend Lisa Doenitz remember my birthday. They bring Jill to my door, a girl friend of theirs, wrapped as a gift tied with yellow satin ribbon.

In my wandering mind there have already been thoughts of going to New York, not as an Art Director but as a photographer. Spending time with Donovan and Duffy I see the grass is considerably greener their side of the fence. They're enjoying much more freedom than advertising affords me. The idea attracts me. I'm already taking photographs of professional quality. Tailor and Cutter recently bought by Clive Labovitch and Michael Heseltine and turned into TOWN Magazine, hired me as a design consultant. While I'm their consultant I photograph several spreads on a men's fashion story and a cover for them. They pay me with a return flight to New York.

I spend a week in New York, totally smitten by its physicality, its music, and its people. I find myself giggling as I walk down its canyons of skyscrapers. The thought to quit advertising and arrive in New York as a photographer starts to fascinate.

Invited to a party at a Knightsbridge model agency I meet Pat

Hughes, a talkative American, attractive in a Chanel suit. We become friends then lovers but at a distance. I don't invite her to move in. She's from New York and talks about many famous photographers and Art Directors she knows. She thinks I could easily get work there. She has a small apartment on Second Avenue. Says I could stay with her until I found something of my own.

88.

Tembe Crossfield Villiers is a beautiful African woman about twenty years old with a handsome, slightly older, English boy friend, Nick Harman, a journalist on the Economist. They're often at the Stein's parties. Jean and I know them as a couple like us and we're quite close with them. One of the striking things about Tembe is her plumb-in-mouth upper-middle class, county accent. I had seen many African faces but had never heard any one of them speak like her. I find it charming.

She comes from a distinguished South African, Xhosa family, her grandfather, Dr Jabavu, was one of the founders of Fort Hare University, the only university for Africans in South Africa where Nelson Mandela and many other leading black South African politicians and writers were educated. The Jabavus also published the only newspaper in South Africa in the Xhosa language.

Tembe's mother, Maree Jabavu, writer and BBC broadcaster, married a Trinidadian airman named Villiers; he was Tembe's father, but died during the war. After the war Maree married Michael Crossfield of the Cadbury family, a wealthy Quaker and filmmaker. A kind and temperate man, he adopted Tembe investing her with some of the more pleasant idiosyncrasies of his class.

I'm wallowing in my own personal despair when Tembe tells me she and Nick have decided to split up. I'm sympathetic but selfishly welcome the news. Here's someone with whom to share my own sad confusion. Sudden impulses invade the mind. Tembe is beautiful and totally distraught by her break-up. Both of us in search of solace, we fall into each other's lives. We seek fun together and become lovers. Neither of us feels particularly suited to the other; we're both dealing with the pain of loss. We enjoy each other's company but there's no sense of permanence. She becomes pregnant.

My friendship with Pat Hughes continues. We now talk constantly about going to America together. There are things that have to be done. I apply for a Green Card, arrange with cousin Peter my divorce from Jean, and put the house on the market. We plan to drive across Europe and around Italy pausing in Rapallo to visit some of her relatives, then all the way down the Mediterranean coast to Calabria, around Sicily and back north along the Adriatic.

Although Jean and I were no longer married, we maintained a fairly close friendship. She married Bill, a successful broadcasting musician and highly regarded ornithologist. After seven years of my marriage to Jean, we were unable to produce children; ironically, when she married Bill, they soon produced two wonderfully intelligent and creative daughters, the elder showed talent in acting and writing, the younger proved to be a consummate dancer and teacher, Often one or other would be included in my post marital meetings with Jean. The girls jointly arranged a party for Jean's eightieth Birthday. Three husbands, past and present were there.

Even just as friends I found her difficult to discuss anything; when I was disagreeing with her about something she would continually accuse me of being obstinate. I always thought I was an open person. Then I considered the death of Jim her third husband. Jim was a lovely guy, an intellectual drinker from Glasgow. His background was much the same as hers and in a way they were very much alike. Maybe poor Jim was the obstinate one when it came to intellectual conversations about life in general. They were both smart and a match for each other. I could see how much she loved him but she couldn't dissuade him from the booze, and that was what did for him. After he died I could see how much he meant to her and she missed him very much. Maybe that's why I found her behaviour, in discussions with me in particular, a little odd, hard edged. We were both the same age, give a month or two, at the tail end of our eighties. Neither of us could believe how old we had become and how the world had changed around us, in many ways to our disliking. I tell this now to better understand later events.

Tembe has her baby, a boy. Her family tradition requires the first-born male to be named Tengo. I visit them in hospital. Another, close friend to Tembe is Matthew, a consummate and well-known typographer, who had been in love with her for some time and now wishes to marry her.

The Green Card comes through; the house sells for a small profit, which I share with Jean. I move temporarily into an apartment in the same block as Terry Donovan in West Hampstead until Pat and I are ready to leave. I fly to Schiphol for a day to the duty free camera shop; I buy a smart black 35mm Nikon with a 2.8 lens and then we're ready to go. We decide to bring the car with us to America so end our Italian tour back in France where we board the SS France – car and all - in Le Havre.

Part 2. The moving finger writes….

89.

When we arrive in New York we find people are still stunned by the assassination of John Kennedy, their bewilderment soon to be overtaken by a hedonistic tide of mind-bending drugs, sex and rock 'n' roll. My first month staying with Pat in her Second Avenue apartment was idyllic. She showed me how to manage in this fascinating city, introducing me, as she said she would, to all the top Art Directors and photographers in the business.

I reconnect with Richard Heimann, a photographer. I'd met Richard, with his agent, in London when they were visiting Advertising Agencies looking for work. Richard, a fine, imaginative photographer, is a hopeless anglophile and very pleased to see me. He immediately offers me the use of his large, well equipped, mid-town studio where I can practice my new art and put together a portfolio. He never comments on my work. Advice, he says, is for passing on. But when I get my very first, albeit unimportant, commission he loans me his Hasselblad (at that time an exclusive and expensive piece of Swedish gear) and acts as my assistant, loading film into the detachable Hasselblad backs while I shoot. He does everything possible to get me started at a professional level.

My relationship with Pat begins to sour. Maybe she resents the amount of time I spend with Richard. Within a few weeks, living together becomes intolerable. She's jealous and angry. She feels insecure, feels I don't need her anymore. I have to leave. Richard has an apartment on Madison Avenue and says I can sleep on his couch until I find a place of my own.

Within a month I've put together a portfolio of reasonable fashion photographs that Richard's agent, Joe Cahill, thinks is viable. He introduces me to Frank Cowan, a well-established New York advertising photographer who'd bought a defunct powerhouse from the Con Edison Electricity Company, a huge industrial space on East Sixth Street between First Avenue and Avenue A. He's looking for another photographer to share the space and help pay off the mortgage.

Frank's highly professional agent, Nob Hovde, flicks through my portfolio making sounds of approval; it isn't so much the work that excites him, it's my English accent. He agrees to represent me and wheels me in front of all the top Art Directors in the city. They think the pictures are more or less O.K. but as Nob had foreseen, they're charmed by the English accent and the commissions follow.

I'm still dossing on Richard's couch and it irks me. He says nothing, it's me that's grown uncomfortable about it and in a whiney sort of way say to him that I'm beginning to feel obliged to him. He answers tersely,

"If you're feeling obliged you'd better get out."

It was unexpected. I thought he'd probably say, "Don't worry about it old chap." and life would go on. Richard was like that. He could see through people's games and gave them no oxygen.

I rent the first floor of a Brownstone on East Sixth Street close by the studio. In my first year, Nob builds me into a successful, very well paid entity. I can afford more cameras and lenses; two Hasselblads another Nikon all neatly arranged in two foam-cushioned aluminium cases with my name stencilled large on their lids. Because of the prices Nob charges for my services, I find I only need work two or three days each week to more than cover the cost of my high life-style, leaving time for me to join my friend Peter Oliver, another English photographer, on his boat for a little Gin & Tonic sailing out of City Island up in the Bronx.

My life has completely changed. I no longer carry the weight of angst that bedevilled my marriage to Jean. I seem to have gained in confidence and can see more clearly what matters and what does not. I used to be closed to anything other than the way I was brought up. I used to believe there was an order of things, a correct way. I'm free of that now and wide open to new experience with little regard for consequences.

I meet beautiful and self-assured Johanna Lawrenson, a model. She becomes my tutor, opening up for me aspects of New York life little known to Pat. She's a young woman of mystery, subtly encouraging me into spaces I've not previously been. She revives my interest in marijuana and leads me into realms of undemanding love.

90.

It's February and bitterly cold, the city beautiful but cold. In spring, summer and fall, midtown Manhattan is hot, dirty and noisy. Heavy traffic negotiating multiple road works, sidewalks crowded six or seven abreast moving at a fair clip. For a few weeks in winter, under a carpet of snow, it becomes an almost silent wonderland. Through the mist of early evening, mid-town skyscrapers twinkle like fairy palaces against dark, snow-filled skies.

A Fifth Avenue store wants me to go somewhere warm to photograph their summer fashion collection, a two-week tour of South American west coast countries in concert with an airline and chain of luxury hotels. They've chosen the model Candice to wear the clothes. I'd seen her pictures in Vogue and Bazaar but never met her, never worked with her.

Before leaving my apartment for the airport I buy a pack of Tampax, empty a couple of the tubes, stuff them with marijuana and return them to their wrappers and packs.

At the airport I watch Candice kissing good-bye the man who'd brought her there. She tells me later he's her analyst and he's in love with her. Without a word, I hand her the Tampax pack to carry through customs. She accepts it without look or question puts it in her bag and hands me a film can. By evening we're in the middle of Fiesta in Panama City, stoned and in love.

Each day for two weeks I adore her beauty through the viewfinder, languorous and solemn, silken brown hair half way down her back, dark, searching eyes sometimes soft as velvet, sometimes hard as jet. She has my complete attention. She says she's in love with me. I'm already crazy about her. In Bogotá I photograph her in the bullring with famous toreador Josalio di Columbia and feel a twinge of possessiveness.

We fly up to Cusco in an old Peruvian Airlines DC3. She has a portable record player on her lap playing the Beatles latest album, Rubber Soul. In Cusco we take a mountain train up to Machu Pichu; more dresses, more photographs. Days of love and wonderment.

When we get back to New York she locks up her fashionable midtown apartment and moves into my one-bedroom pad on East 6th Street. She likes the idea of slumming it for love.

She claims to be part Apache. Street-wise, she comes from nowhere and despite a complete lack of formal education, makes her way with ease and confidence like she was born to it amongst people of fame, wealth and influence. Her beauty, her smouldering sexuality is irresistible, the way she looks at you, a shadow of a frown, a slight purse of the lips, a smile somewhere between amusement and pity.

She's the first woman I've lived with since my divorce and I'm getting to revisit some of the advantages and disadvantages. She makes the place her own, clothes, underwear, stockings left lying around, table lamps draped with silk scarves, cut flowers imported daily. I don't mind this kind of feminine disorder,

She likes to be seen with me. We eat out every night in fashionable restaurants. She buys things on a whim and quickly tires of them like the New World monkey that flew around the apartment for a few days tearing the place apart, shitting everywhere until she donated it to the New York zoo.

Her sexuality is a quicksand in which I'm happily sinking. Apart from eating, we only go out when invited to the parties of famous friends of hers like Hunt Hartford or Salvador Dali. Most evenings we're home making love. In love she's a whore. She has many aids from silk scarves to cocaine and amyl nitrate. She ties me to the bed and sometimes invites a girl friend to come and play with us. I'd never before experienced anything like it. It deepened my fascination for her

There are occasional break-ups, lovers' spats when she'd go back to her mid-town apartment in a huff. She gets bored with happiness, picks a quarrel, inventing insane jealousy of a model I photographed several times that week, accusing me of fucking her. I swear it's not true but she drives herself into a fury, packs up and leaves.

During one of these separations, I'm with a new friend, Bob Mayo, playing extras in a crowd scene in a Larry Rivers movie, a party on a boat on the Hudson. I meet Alex. She and her friend, Viva, one of Warhol's stars, are amongst the crowd. Bob goes for the star and they leave. Alex and I end the evening at my place.

We're in bed. The phone rings. It's Candice. She says if I don't kick the bitch out right now, she's coming down to kill the both of us. How did she know? Alex dresses quickly. Candice moves back the following evening. The incident has in some strange way promoted me in her esteem and she's very loving.

91.

I thoroughly enjoy the life of a fashion photographer. I'm extraordinarily satisfied creating images of beautiful women for a living; but I'm not impressed by the protocols of the business world surrounding it and, sometimes to Nob's concern, pay it no lip service.

One of Frank's clients is Al Vandenberg, an art director for DD&B, the most creative ad agency in town that every photographer wants to work for. He's often at the studio working with Frank and sometimes drops in to watch me. I dislike being watched while I work but he's from DD&B.

Al's a tall guy with a stoop, a gangling sort of man dressed like everyone else in the advertising business in those days in a three-button, drab–coloured suit, trousers short of a length floating over heavy-soled black brogues. It turns out he's not interested in my work at all. Between shots while I'm changing film, he drones on about himself. He's telling me he's a photographer too, a part time student of Brodovitch.

OK.

A purist, he says. One camera, one lens, one film.

Uhuh.

Already has a couple of pictures in the Museum of Modern Art.

I'm beginning to dislike him. There's an arrogance about him. He's dismissive of most of the commercial work he sees around. I guess that includes mine. "A photograph has to tell it like it is. Anything less is a jerk-off." He lowers his voice. He's quitting DD&B and moving to England.

Yeah?

He's told I'm from London. Can I give him names of people he can hit on for photographic work when he gets there?

No problem. I'll be glad to. I write down one or two names and addresses and wish him bon voyage. On the day of his departure for the Old World he calls me and says before he leaves there's a guy he wants me to meet, someone I'd really get along with. He was insistent.

Perhaps you know what that's like? A virtual stranger, for whom you have no particular affection, presuming to choose friends for you? Reluctantly, I agree to meet him at the Ninth Circle, a popular bar in the West Village, five minutes from where I live.

He shows up with this tall guy, dark haired with startling blue eyes. A young woman he knows at the bar immediately distracts Al's attention and he leaves his friend to introduce himself. Gazing into those blue eyes I shake hands with Bob Mayo. For no reason I can think of I feel totally at ease with him like I've known him for a hundred years. It happens like that sometimes.

People destined to become friends often meet each other by the most unexpected routes, their combined fates working away at it even as they sleep. If you look back, every event in the life can be construed as another inevitable step towards where you're standing now. Al says his farewells and slopes off to catch a plane to London.

92.

Bob. Now there's a story. He was a cop in the Bronx when he fell in love with a young woman, Helen, he saw regularly on his beat. He'd be strolling out of the precinct at precisely 8.30 a.m. and she'd be waiting at the bus stop for the 8.35 to take her to her job in Manhattan. It started with morning greetings and grew into a hot love affair.

Helen, when her job sends her back to San Francisco where she comes from, Bob turns in his gun, badge, wife and kids and follows her. He divorces his current wife and marries Helen but the relationship didn't work out. After a year or so he returned chastened to New York.

He's not a cop anymore; he's an artist, a painter living off the GI Bill renting an Artist In Residence loft down on Grand Street, down in SoHo (South of Houstan). He sees his kids from time to time but there's no going back to the wife. Not that she has any axe to grind. She'd been happy about the divorce and took the opportunity to marry her boss who'd been trying forever to get into her pants.. She and the kids are living in style and comfort in Westchester.

Bob and I spend a lot of hours together. We eat, go to movies, blow a few joints, listen to jazz, Miles at the Vanguard, Monk at the 5 Spot. We hang out all hours sorting the Universe. LSD is legal and we're experimenting with it enjoying the discovery of new perspectives; there's a lot to talk about, reassessing some of the things we'd taken for granted all our lives.

A New York filmmaker we know who'd been travelling the Indian sub-continent making an obscure documentary on Hinduism, brings back with him to New York a Swami he met while filming in Sri Lanka.. He set him up in an apartment midtown and invited a few people to meet him.

Dressed in a floor-length saffron-coloured robe, he's a tall, slim, handsome fellow of indeterminate age, a full head of black hair long to his shoulders and equally long, greying beard. Dark eyes shine from a healthy, smiling face. He sits, cross-legged, and introduces himself. Satchidananda. He's a Yoga Raj and would be happy to teach us what he knows.

Neither Bob nor I were easy about submitting to spiritual disciplines beyond learning a few Yoga *asanas* but we're comfortable with him. We relate to him as a personal friend and he responds to us in that way. We take him to the movies to see a Hollywood spiritual blockbuster. One weekend the three of us drive up to Niagara Falls. He loves driving and takes a turn at the wheel. The trip turns out to be auspicious; there are no Waterfalls this particular weekend. They'd been diverted for purposes of maintenance of the rock face.

Swami's *au fait* with everything going on, totally at ease with the sophistications of the modern world. He wasn't always a Monk, only since the death of his wife. Until then he had an electrical business. Our friendship isn't destined to be a close affair - we occupy such a small part of each other's lives but his very existence becomes important to me. I'd not met anyone so calm, so perfectly balanced.

Candice gets on well enough with Bob, fascinated by his forensic stories of cops and villains. Bob and his cop stories. His attraction to her is plain to see. When you have Candice on your arm you can expect attention. I don't hold it against him. I love him but there's no way I'm going to share Candice with him or anyone else. Candice is fiercely loyal; she'd leave me before she'd sleep with someone else.

In the spring of '66 one of the Fifth Avenue stores commissions me to cover the Paris collections for their fashion department. I use Candice as the model. We're packed and ready to go. Bob visits to say goodbye. He's uncharacteristically downbeat; feeling left out. Candice in her nonchalant matter of fact way tells him to dry his tears and go pack a bag. He protests he doesn't have a nickel. She'll cover for the trip.

In Paris we stay at the *George Cinq* for the duration of the Collection madness. As soon as it's over I hand the bag of films to the store's buyer to take back to the New York studio for processing. We're staying on for a few days; we'll see her later.

Candice, Bob and I move to a charming slum of a hotel on the Rue de Seine where, according to Candice, the Beats Kerouac, Ginsburg and Burroughs used to stay and that's where Bob meets and falls in love with Kelly, another successful model from New York. She's finished the collections and is on her way to London where she's signed with a top agency. She's crazy about Bob and she's taking him with her.

I'm pleased he's found a mate but sad our individual involvements are drawing us apart. We've known each other such a short time but are closer than blood brothers. Candice throws a farewell party for them, dinner at La Coupole. That's how she can be at times, romantic, generous to a fault, in tears at the departure of such dear friends of only a few weeks.

After Bob and Kelly leave, Candice decides we're not going back to New York at all. I didn't give it a second thought. The whole of the past, the New York apartment, all my possessions, the studio, Frank and Nob, clients, everything had lost its importance. We rent an apartment on the Rue du Bac in St Germaine and are so involved with each other time passes unnoticed.

I telephoned Richard Heimann in New York begging him to send my personal effects, which amounted to a few bits of clothing, my portfolio and my car, the MGB. He raided my apartment but someone had beaten him to it and none of my personal stuff, nor my portfolio was there, but he did manage to send the car, which duly arrived at Le Havre. I went down there to pick it up and drive it back to Paris. It gave our lives another dimension; we could visit places otherwise too difficult or too tiresome to visit. For a time, it staved off the inevitable.

93.

One morning there's a perceptible change in her mood, she's bored again. When things go too smoothly, too predictably, she begins to get edgy, begins thinking something must be wrong. Perhaps, deep down, she feels she doesn't deserve to be happy and before you know it she's not happy at all. I try to get her to lighten up and she replies in her casual, sarcastic, Californian way, "What makes you think *up*'s the place to be?" By the end of the day she's morose. "I'm going downtown" she says, her voice already sluggish from Vodka, hashish, opiates.

I follow her. I want to be wherever she is, go wherever she goes. Maybe she's right. Maybe downtown has fascinations I've yet to discover. I'm in love and don't want the romance to end. After a week we're taking so many drugs and booze I can't tell night from day. I fall into unconsciousness not caring.

Awake again; hung over and ornery, we're at a friend's studio to score some marijuana. He's doing a job for Elle magazine. We're waiting in the gallery overlooking the studio impatient for him to finish.

Whose turn is it to pay for the dope? I've had no income since leaving New York and money is beginning to be a problem. An argument flares, incriminations, blame, threats. Our friend is trying to deal with clients and models down in the studio while overhead we're yelling at each other and throwing furniture. Everyone looks up at us with amazement; someone suggests calling the police.

A few days later I call an end to it. She smiles that smile and shrugs. She has a wry sense of humour. The only time I ever saw her laugh, really laugh uncontrollably like a kid, was after an orgasm.

94.

I'm emotionally drained and unsure of the next move. New York is no longer an option. Frank and Nob freaked out when I didn't return and that's all over. I've traded professional success and financial security for love and I'm back on the pavement.

It's in my nature to always be moving towards something, to get something to happen no matter what. I drive back to London and crash with Bob and Kelly. They have a neat apartment in Kensington and a spare room. While staying with them I meet Barry Hall, his wife Jacqui and their two infant sons. They're English friends of Bob. They knew each other when he was living in San Francisco with his second wife, Helen, the one he met when he was a cop in the Bronx.

It only happens a few times in the life you encounter someone who brings with them new information, a new perspective. Meeting Bob was significant. He and I are carved from the same rock. We came into each other's lives for the fun of it no questions asked. Barry is a darker soul. Meeting him opened doors for me into the obscure, introducing me to possibilities I'd never before considered, broadening my concepts, smoothing away my shibboleths. Bob and I recognised each other as soul mates. Barry, although younger, was, for me, a teacher.

Neatly trimmed beard, good-looking, modestly elegant, he, Jacqui and the children live in a second floor flat in West Hampstead. Jacqui, bright and attractive, concerns herself with the practical things in life, bringing up the kids, seeing there's food in the house, paying the bills. They're an inquisitive, creative, congenial couple. We become friends without question.

Barry publishes poetry in a converted mews stable around the corner from the flat, mostly the work of poets he knew and admired when he lived in San Francisco. Alan Ginsburg, Dianne de Prima, Michael McClure, Robert Creeley, Charles Olsen. He makes all his books by hand, designs cover and page, sets type the traditional way in a compositor's stick, prints it on an old cast-iron Eagle press. His books are gems of patient craftsmanship and have already become collectors' items. He's considered by the Times Literary Supplement to be the world's finest publisher of contemporary American verse.

But despite the accolade and reputation, he and Jacqui are broke. There's no money in poetry. The business barely pays the rent. He's finding fulfilment and recognition but he's cash poor and uncomfortable with it. It cramps his style.

I could see it the first time we met. Behind the practical publisher lurks a man of secret dreams and visions. He doesn't talk about them, rarely shows any emotion at all other than an occasional wry smile, a chuckle into his beard. But beneath his charm I can see he wants to be more than editor and publisher. He wants to be the artist, he wants to be the author, wants to indulge *his* fantasies, make real *his* dreams. It matters to him more than he lets on.

95.

Looking for a way back into the photo business I meet Roy Pegram, a photographer's agent who's impressed by my New York experience and reputation and wants to represent me.

My cameras have been gathering dust and I'm broke since the expensive spin with Candice. Roy finds me a studio on Flood Street in Chelsea that used to belong to the painter Augustus John, a magnificent space with a mezzanine bedroom, a kitchen and shower out back. He supplies me with plenty of electronic flash and rolls of no-seam paper. Within days he's bringing me sometimes interesting but always well-paid work. I'm back in rhythm.

I get a phone call from Alex, the young woman who'd fled from my bed the night in Manhattan when Candice threatened to kill the both of us. She's in London to model and would like to show me her portfolio.

Discovering Candice is out of the frame she lets me know with a degree of pathos she has nowhere to stay. Candice may no longer be in my life but her ghost is still around. I miss the warmth and smell of her silky smooth body and the intimacies we shared. I don't find Alex attractive in that way but she's in need of kindness so I invite her to stay.

She was two months old when her mother walked out leaving her with dad, an artist and intercontinental traveller. He'd been machine gunned in the Philippines in world war two and now lived well off a substantial disability pension. He'd be back and forth between Europe and the US; a couple of years painting here, a couple there, then back to New York for an exhibition. She became part of his baggage, growing up in America, England and Italy amongst his arty associates and friends, step mothered now and then by his various lovers. They lived well, in aristocratic style, but it didn't do much for the child's self-confidence. Impermanence. No sense of home. No continuity, never knowing what to expect next. Whatever the reason, she's a quiet girl, easy to live with, reluctant to join in trivial chitchat. In general she doesn't have much to say or perhaps she has so much to say she doesn't know where to begin.

I'm a daylight photographer. I know how to use flash but I prefer the soft, forgiving light of day. It floods into the studio through a large window in the pitch of the ceiling but at too high an angle to give good light to anyone on the floor. To get the angle I want, I raise the floor. I build a platform four feet off the ground and that's where everything happens, on the platform. Even if you're just visiting, I'll get you up there and shoot a few Polaroids for the hell of it. The walls around the alcove at the back of the studio, where clients sit watching me work, where we sit of an evening with friends, are covered with these shiny little black and white Polaroid prints, faces from all the highways and byways of life peering out at you, distinguished, undistinguished, young and old, male, female, an anthropological document you could say, a Polaroid of the human race.

The place is an energy centre, always something going on. I can't account for it. Dissident South American poets, revolutionaries from Paris, anarchistic space cadets, all find their way here, all of them protesters, all seekers after some kind of truth. I don't know why they think they'll find it here, but I enjoy their company. They're interesting people. We talk, exchange ideas and I shoot Polaroids of them.

Most days of the week the place is a hive of industry, stylists fussing around models on the no-seam, messengers coming and going, phones ringing, magazine and advertising clients with their problems and anxieties. Roy keeps them off my back. He's an elegant, well-bred drunk who can really knock it back no matter the time of day. He says his drinking is down to a traumatic experience in the war that left him terrified of everything and drunk is the only way he can deal with the world. I find no fault with him. He keeps the clients cool and brings in the work.

I still can't get Candice out of my mind. I find myself groping blindly for a way back into that sensuality of the past, treating Alex to some of Candice's preferences, taking her to expensive restaurants, buying her clothes and perfumes hoping, perhaps, she'd metamorphose. But it's not happening.

96.

I'm at Barry and Jacqui's. It's the evening the news comes down the grapevine about Neal Cassady. Neal was Jack Kerouac's alter ego, Dean Moriarty in Jack's novel *On The Road*, the legendary driver who drove him back and forth across the States in search of a father he never found.

Neal drifted over to Oregon to hang out with Ken Keasey and his Merry Pranksters, driving them in their psychedelic school bus every which way across the States enlightening the good folk of America about sex, drugs and rock and roll. The news was he'd gone missing a few days ago and had just been found on some railway spur in Sonora, Mexico, dead of drugs and exhaustion age forty-one.

The Twentieth, as well as being atomic, is undoubtedly the century of the automobile and Neal was the archetypal Twentieth Century American driver, the car a seamless extension of his consciousness and sensibilities. He had technique. I heard he could reverse park, be slamming the door and walking away before the car had come to a neat halt between two others.

Our minds float nostalgically around this culture, all the driving stories either of us know rising with our hashish smoke. Barry recalls a news item he'd read in the San Francisco Chronicle when he'd lived there, an insignificant piece, a couple of column inches.

A life of unintended consequences by Malcolm Hart

The California State Highway Patrol flags down a car on account of some minor infringement of the law, a broken taillight or something. The driver, instead of stopping, puts his foot down and speeds away.

Thinking they've happened upon a felon, the cops give chase. They run him off the road at great speed and he's killed. The subsequent enquiry reveals he had no criminal record, isn't involved in crime at all. Like the cops who pursued him, he liked driving fast, enjoyed a chase.

It's epic. Freedom challenged, fought for and paid for with a life. I can't get the story out of my mind, its inevitability. I can see the movie, its beginning, middle and end. Every evening after work at the studio, I write the story, a drama triggered by innocent events that gain momentum on an inevitable course like a Greek tragedy, fate portended. I call it *Pick A Card, Any Card*. I show it to two producers, one of them a friend I know. They both want it. My friend buys it on the spot.

Barry.

A life of unintended consequences by Malcolm Hart

Bob & Kelly's Wedding

A life of unintended consequences by Malcolm Hart

Jean Hart

97.

Saturday night is party night at the studio. Bob and Kelly, Barry and Jacqui and a few others ready to relax. I rent a 16mm projector and movies, the more esoteric kind, films I'd heard about but not yet seen made by artists like Cocteau, Bunuel, Dali, Jean Genet; films by New Yorkers Lionel Rogosin, Shirley Clark, Jonas Meekas. Like painters at a canvas they all make films to please themselves with no one else to answer to.

After the movie there's music, maybe some dancing, a few bottles of wine, something to smoke. These are good times and we all look forward to Saturday night at the studio. Word gets out and acquaintances I haven't seen for years start showing up as if by accident, just passing by, thought they'd drop in. Each weekend more and more people come to the door and introduce themselves as friends of friends.

I'm amused at first then concerned the way our private Saturday nights are becoming a public affair and I could think of no way to stop it. When the pubs empty on Saturday night it seems like the entire population of Chelsea, pissed out of its mind, is drifting down Flood Street to the studio.

A drunken fight brakes out on the stage, people punching, jostling, angering those around them who join in. A riot; complete strangers, men and women, screaming and kicking. The stage shudders and gives way. Neighbours call the police.

It was a sign. The studio was no longer the place for me. I could say that's what changed the course of my life but it was already in my mind to give it up. I'd been feeling edgy these last weeks, no longer sure fashion photography was the greatest adventure life had to offer. I used to wake of a morning charged with energy and anticipation, keen to get the day under way. It went missing about a week ago. It didn't happen. I felt lazy and uncharacteristically moody, everything seemed too much trouble.

Alex is in a downbeat mood. She thinks my *ennui* has something to do with her; maybe there's a shadow of truth there. My attempt to mould her into being more like Candice is a failure. She doesn't come anywhere near the fantasy still teasing my mind. She's too young. She lacks the unpredictability, the humour, the sly sexual aggression. She knows I'm not crazy about her in that way and it makes her nervous. Adding to her angst, her looks don't conform to this year's Vogue paradigm of beauty; consequent failure to get work leaves her depressed doubting her self-worth.

My *ennui* has nothing to do with her. My life as a photographer no longer satisfies me. Something has been lost. Attempting to record the precious moment in its frozen perfection isn't doing it for me anymore. I'm beginning to resent the habit of photographing everything; a momentary glance from a pretty girl, an early morning landscape, spectacular displays by the sun. I'm beginning to resent the camera, constantly between my eye and whatever it is I'm looking at, leaving no opportunity to contemplate. It's time to pause, step back, and check out my options and breath deeply.

I have changed. A year and a half with Candice changed me. Throwing my career and the life surrounding it to the wind changed me. No one died but everything changed. The planet still spins on its axis. I'm free to do what I like. What does it mean, 'free'? What is it I like?

I have no plan of action. I don't know where I'm going but confident as an idiot I'll get somewhere. I have money from the sale of *Pick A Card*, I can afford to take my time and wait for destiny to reveal itself. Alex and I go to Kathmandu.

98.

I'd not yet visited the Indian subcontinent, the source of so many of the spiritual notions that continually drift through my mind. In Nepal I witness a simple way of life embedded in basic values, a haven of peace and gentle industry distant from the harsh competition for survival in the west. Life is simple in Nepal, a well-nourished, well clothed, well housed, low-technology society, easy for the mind to manage. There's nothing pretentious or self-important about them. They don't laud their artists. No one knows the names of the architects who designed their houses; no one knows who painted the murals in their temples. They're not signed. They're not sanctimonious either, not even in their Buddhism. They drape laundry to dry on holy stupas; their dogs piss on sacred shrines.

99.

We're back in Chelsea staying in the elegant Glebe Place home of one of Alex's father's ex lovers, one of her stepmothers, while we look for somewhere of our own. We're not specifically looking at this moment, not studying estate agents' windows. It's our first week back from Nepal and the mind needs to readjust. We're simply strolling down the Kings Road in the warm autumn sunshine readjusting the mind when we bump into Maree.

Maree's a freelance stylist in the photo trade; she's worked for me on a number of occasions. I like her. Alex likes her. We're happy to be bumping into her. We sit outside at the Picasso in the sunshine drinking coffee, chatting warmly about nothing in particular, and it just slides out along with the rest of Maree's gossip.

She and her partner, a TV commercials producer, have just moved into an enormous mansion flat in Fulham and are looking for someone to share the space. We go to see the flat that evening and meet Maree's partner. We enter a large room with white walls and ceiling, polished wood floor and tall, casement windows opening onto a balcony overlooking the street. Alex is impressed.

Mike Margetts struggles up from deep in a black leather and chromed-steel armchair, with its pair the only furniture in the room. He struggles up with drink in hand and greets us, smiling. He's six foot two, well-built, English good looking, wearing a suit.

Maree and Alex disappear to update their trade gossip. Mike offers me a drink but I'm not drinking at this time. I roll a joint. Some folk would take offence when I rolled up in front of them not only because it's illegal. Mike's not offended, not at all. He's curious.

I light up, smoke a little and pass it to him. He accepts. I can see he's not a regular user. We smoke and talk and soon he's confessing to not being totally happy with the career path he's chosen. As a producer of commercials he has to interface daily with impersonal advertising agents and businessmen. He's finding the corporate aspects of the production business predictable and boring. He used to be a cameraman and editor, part of a creative team and now he's a suit because the money prospects are better. He misses active involvement in the film making process. Excitement has gone missing from his life.

He stares unseeing at the joint held between his thumb and forefinger. He sighs, genuinely sad, thinking perhaps he took a wrong turn. There's not much I can say. I know how he feels. I shrug. He grins and I like him for it. He and Maree give us a quick tour of the flat. I get the nod from Alex and tell them we'd like to share with them. Everyone's happy. We make a deal, I write a cheque, shake hands and arrange to move in the following day. Mike hasn't commented on the smoke but he's smiling a lot.

As we're leaving I notice a stack of 16mm film cans on the floor behind the door, the only things in the naked room besides the steel and leather chairs. I ask Mike what's in them, thinking they were commercials he'd made.

He says "Out of date virgin stock. Unexposed film."

I say, "Let's expose it."

Still smiling, he says, "OK."

A few days after we move in, he and I start to make a movie. This is it. I am suddenly an irrationally confident filmmaker. I've found it. This is what I do. I am a filmmaker.

Mike is Director Of Photography; I'm the Director. Inspired by those filmmakers whose work I used to show Saturday nights I avoid doing anything I've seen before; no script, nothing. No script, no storyboards. I want to see what happens if we predetermine nothing, no concept to hold us steady, just our instincts and sensibilities to rely on. No judgmental stuff like I often see in movies, no messages.

Mike's a perfect partner, experienced in all aspects of film production and ready to explore new paths. He's owed favours up and down Wardour Street and borrows a 16mm Arriflex with a three-lens turret from a facilities company he does business with. He opens the first can of film. As he unwraps the film from its black, lightproof paper he holds it under his nose and breaths in deeply, smiling as though stoned on the acetate smell of it. He threads it effortlessly into the camera.

I have a piece in mind, an erotic meandering. A virgin pursued, who, finally cornered, metamorphoses before your very eyes into a seductive, inviting whore. Sublimation? So what? It's a nice story, no message, nothing difficult to grasp and a sense of completeness about it.

Mike likes it. Alex likes it and agrees to act the role. The first shoot is in woodland outside London. The three of us drive there crammed in my MGB.

Mike chases Alex filming her on the run, pale without make-up, wearing a long, leather coat. Dashing through the woods she pauses now and then to catch breath and cast anxious glances behind her. She's a frightened virgin being pursued. We do this for an hour or two then drive home dropping off the film at the lab.

Next day we race to the lab to see the rushes. Alex is good. Exposure, focus and colour are good. Mike still has commercial commitments to meet so it's some days before we can shoot the next scene, the metamorphosis from virgin pursued to seductive whore.

We drive to the Suffolk coast to a place where forest ends and beach begins. Alex transforms her face, draws heavy eye make-up, paints a lurid gash of a mouth. She's wearing the same long leather coat but little under it. Bare breasts, garter belt, stockings, some weird chains, jewellery and stuff.

Camera running, Mike chases her out of the woods onto the beach. From behind, she looks exactly as she'd looked in the previous shoot. Now, cut off by the sea from further escape, she stops and turns slowly to camera. Instead of the frightened virgin we've come to expect, she's smiling slyly like the painted whore she's become. Her coat falls open to reveal her lovely breasts and body draped in this weird regalia. She sinks slowly to the sand, smiles and smiles, inviting us down. Mike follows her down, coming in close. Cut. OK.

It's so cold we only do the one take, pack up and drive home. The following day we go to the lab's viewing room to see all our rushes projected. There's Alex running scared through the wood out onto the beach and, trapped, turning into a delicious seductress. We're on our way.

We are turning the camera on things we'd only thought about photographing. Mike always wanted to shoot the sun. He captures it on a 600 mm lens as it sets, the source of all known light, a wonderful orange and scarlet fiery disc smouldering at the edges creeping into frame against a purplish sky until it more than fills the screen.

A friend, American photographer David Gittens, had designed and built the quintessential sleek and sinister car, a design way ahead of its time. Tapered front, high back, no doors. To get in, the whole roof and windscreen of darkened glass, hinged at the front opens in one. It's a powerful image. We were to film its virgin run starting outside the pillared and porticoed stately home Maree found for us somewhere in the Berkshire countryside.

It's a raw January afternoon. The car waits at the foot of the stone steps, Alex stands under the portico at the top of the steps, veiled and mysterious. Mike films her walking down the steps to the car. The stone is icy. He's walking backwards down them as he films. He slips, falls flat on his back, camera held high. Cameramen are heroes.

They could be hurtling off cliffs to their deaths and all they can think of is the camera and the precious film. I go to help him. He smiles up at me, his trousers all rucked up, I notice he only has one leg, one real one. The other is pale pink plastic, a sock and boot on the end, of course. You'd never know it by the way he walks, never a suggestion of a limp. He ignores my offered hand and struggles to his feet, brushes himself down. I ask him about the leg. He says a bus ran over it when he was a kid. He was kneeling at the curb fixing his bike.

The film has invaded our lives. In between Mike's commercial commitments, we shoot all sorts of stuff, anything that strikes our fancy.

Not thinking about a movie or an audience, blindly confident that anything that interests us for whatever reason is worth a few feet of film that might finally find its way into a movie we haven't yet formulated. I've no idea where these disparate bits and pieces, these images and ideas we've produced, will end up in the film. I've not given it a thought.

We need a sound track. The movie's obviously playing in an erotic scale. We have a sinister motorcar; we have sex. Willie's an American actress playing Marilyn Monroe in The Beard, a surreal piece by American poet Michael McClure at the Royal Court. I know the director. Mike and I go to see the play and afterwards party with the cast. I corner Willie, tell her about the film and she agrees to take part.

I'm driving the MGB wearing a throat mike. Willie is sitting next to me wrapped head to toe in a blanket, an anonymous bundle. She's wearing a throat mike. There's another mike gaffer-taped somewhere under the dashboard to pick up engine noise. Mike is jammed in the space behind the seats with a Nagra recording deck and a small mixing board. The Nagra's running and Mike is mixing as we hurtle up and down the newly opened M4 motorway. There's no script to refer to, none of us knows what is supposed to happen. Willie starts moaning, wriggling a bit under the blanket. When I change down a gear she responds, a sharp intake of breath as if I'm touching a sensitive place. I raise the revs and she comes with me, sighing as I slip back into top.

Willie and I are beginning to understand what is happening. We do it over and over. One more run. Willie gets into it, rocking under her blanket, making noises like she's really coming. I put my foot down; the engine noise rises to a whine. Willie's cries become sharper, letting go, gasping as I slam into overdrive. The speedometer is hitting it's maximum as she finally comes, screaming, hiccoughing and spluttering. Mike's delighted, yelling encouragement. What a fucking sound track!

We don't have to think about the sequencing of the material, everything fits. Every frame we've shot finds its place somewhere on the twelve minute sound track of Willie's orgasm, rising and falling with it, sometimes startling, explosive, sometimes slow and fluid. It's a poem, a series of erotic implications that have somehow become coherent. There's a beginning, middle and an end. Neither of us can believe what we've done, not even sure what it is we've done.

Maree consults the I Ching, the Book of Changes, an ancient Chinese oracle. People did this kind of thing in the 60s when curious about the situation they find themselves in. It's like going to a fortune-teller seeking a clue as to what fate has in store. I keep an open mind. The I Ching points to a hexagram, a chapter called Arousing. It talks about creative energy, prognostications of greatness and success. We can live with that. We call the film *The Arousing*. Mike gives up his job.

We apply for certification from the Lord Chamberlain's office for its theatrical release in Britain. He turns us down. Nothing he can actually put his finger on, no identifiable genitalia, nothing like that. His Lordship couldn't advise where or what we might cut to make it in his opinion fit for public viewing.

In the places we're likely to get distribution, a refusal from His Lordship is like a rave review. We get an agent accustomed to handling short films. He sells 35mm theatrical rights in Canada and Sweden. A Yugoslav short film festival gives it the Tito Prize setting it up for a long run on the university circuit.

We can't believe it. This mishmash of ideas, this self-indulgence of images and sounds, is being acknowledged, responded to by audiences. It's beginning to dawn on me. More than a sense of achievement, the feeling that something outside of me had been contacted. Is Mike going through the same changes? Intellect, polemics, dialectics didn't come into it. We took the leap, surrendered to the muse and we're being applauded. We can do anything. Revolutionise the industry! Bring truth and grace to Hollywood!

Mike, Alex and I board a plane to New York, The Arousing under our arm. Maree's not coming. She and Mike haven't been getting any further in their relationship and another man's taking an interest in her. I don't know how Mike feels about it. He smiles most of the time so I figure everything on balance must be OK. He's just had his mind blown by *The Arousing* and allowed the course of his life to be dramatically changed. It takes courage to make life changes at the drop of a hat, to recognise the creative flow and go with it. Mike is not afraid. He's enjoying a new sense of freedom and about to see New York for the first time.

For me, returning there is revisiting an old lover. New York and I had a good time and some extraordinary experiences,. It's where the story started, where I met Bob, the Swami, Candice, Alex.

London's a shy place where new ideas take root tentatively, almost apologetically. New York on the other hand, hip to everything creative going on in the Universe, heralds in the new at the top of its media voice, celebrating the outrageous in every creative field. New York lives life in the nick of time. You hear it on the radio every day. "It's a fine morning downtown Manhattan. 8:45 and later than it's ever been!"

We're showing The Arousing to Alan Douglas in his office five floors up on the West side of Manhattan. I'm not sure why we're there; we like showing the film so I guess we're there looking for distribution but we'd show it to him anyway. I knew Alan from the days when I'd lived there. I'd not had much to say to him then but knew him to be an enlightened entrepreneur in the publishing business. He plays his cards face up.

Jazz music is his primary love, producing musicians of talent as yet unrecognised by the majors. He relates to artists, writers and filmmakers in the same way always turned on by originality. He's congenial; pleased to see us. He rolls a joint.

Alan's an elegant, slim figure of a man neatly dressed in pressed jeans and carefully tailored doeskin jacket, Indian-style with fringes. I'm listening to his gravelly New York voice; his heavy lidded steely blue eyes are on me, unblinking. He talks quickly like he's got it all worked out and in a hurry to get on with it.

Blinds are lowered, lights switched off. We run the film. When the lights go up he doesn't say anything, just looks at us and nods. He telephones a friend, a music producer, Michael Lang, who's putting together an outdoor rock and roll concert and wants it filmed.

Michael Lang can't be more than twenty. We drop the blinds in his midtown Lexington office, turn off the lights, run the film. He sits through it in silence. He doesn't even say anything when the lights are turned on. He tears off a piece of wrapping paper from an opened parcel on his desk and with a magic marker scrawls a contract on it for us to shoot his upcoming event, a weekend of peace and music for an audience of twenty thousand.

Mike and I really wanted to jump about and shout but we sit there casually while he gives us a rundown of the bands he's hired, some of the most famous in the world. He gives us money to rent a car, equipment and stock. He wants us to go to the site, a short way upstate, and start filming the crews constructing the stage and preparing the meadow where the audience will camp out. We leave his office in a daze, floating out onto bustling Lexington Avenue, warm air, bright blue sky and the taste of success in the mouth.

The Arousing quickly established us as bona fide independent filmmakers in New York taking us through numbers of different doors and meetings with other creative people. We show it to Michael Wadley's production company. He had been shooting music performances on split screen. He shows us a clip they've just done of Aretha Franklin. It's very professional. We sense a rapport, tell him about our commission and suggest he covers the stage for us at the concert. He's keen. We'll stay in touch.

Alex and I are not there when Mike meets Barbara. I don't know where they met. A party? In a bar? The park? Mike's not shy. Barbara's a tough, attractive little Jewish girl from Queens and Mike's fallen in love with her. She's with us when we drive upstate, trunk full of equipment and raw stock. They're in the back already thick as thieves. I'm driving, Alex close beside me purring like a superstar on location. My mind's full of the film we're going to make, another *Arousing*, an epic poem of the event, the people, human enjoyment cut to the finest rock and roll in the world.

For the next two weeks Mike and I wander the concert site, hanging out with the construction crews, hard hats over their long hair, filming and recording them constructing the stage and its massive towers for the lights that will illuminate the bands. We film a moonlighting Chief of Police, hired for the event as a security consultant. He's on the telephone to a drug company ordering thousands of doses of Thorozene to calm down his policeman's fantasy-fear of thousands of people crazy on LSD. We follow Michael overseeing his project on horseback, riding around the site of an evening attending to flagging energies, a young man with a million dollars on the line.

He's hired a communal family of hippies called the Hog Farm, twenty or so youngsters who'd absconded from their middle-class homes and families. They're up here from their farm in New Mexico to police the weekend's drug behaviour. They're experienced trippers who know how to calm down drug-induced paranoia without the use of chemicals. We spend most of our time with them, smoking their dope, shooting footage of their daily life, working, cooking, recreating. In the movie in my mind I see them as the core spirit and energy of the event, hosts to thousands.

In the last couple of days there have been radio reports of more people converging on us than expected. One report says hundreds of thousands. Michael's amazed. He thought he'd get an audience of about twenty thousand, now it looks like it's going to be upwards of a quarter of a million. Warner Brothers are on the phone to him about film rights. The crowds are beginning to arrive.

Mike and I try to add it all up, its significance to us. Warner Brothers? Global distribution? We're too busy to fantasise. We call the New York company we'd spoken to about shooting the stage. They're on their way.

On the opening day of the show the crowds are so large they've ploughed down the fences. Nobody's paying to come in. We film Brinks armoured cars waiting to take away vast amounts of cash, standing empty. We keep shooting right up to the evening of the first performance. There's still no sign of the New York crew. We're filming cars arriving, headlights lighting up the people around them, hundreds of them milling around in the swirling red dust. Ritchie Havens is coming to the lighted stage and the biggest audience of his life now swollen to near half a million when the New York company finally arrives. Twelve crews all carrying state-of-the-art Éclairs with large zoom lenses, each camera linked umbilically to a soundman. I'm impressed and at the same time puzzled; something's not quite right. Twelve crews? I'm expecting one cameraman, maybe two. I chase around looking for their producer to find out what's happening. When I find him he hasn't got time to talk to me. They're not working with us; they've made their own deal with Warner Brothers. It's not our movie any more. It's their movie now.

We had a contract made in good faith with Michael but reality had changed around it. That's the way it goes sometimes. No one was to blame. Michael couldn't have anticipated the numbers that aroused Warner Brother's interest. Mike switches off the old Arriflex. They'll need our footage; we'll get credit, some consolation. We'd had a brush with fame and fortune, got close but no cigar. We joined the crowd.

Late on the Monday morning after Jimi Hendricks had closed the show with his personal take on The Star Spangled Banner, Michael Wadley approached us asking for the film we'd shot. Mike asked him what was the deal. We had, after all, been contracted to make the film which was now his. He offered us standard Cameraman wages of $125 a day. Mike told him to get fucked. Michael Lang sold the film for us directly to Warner Brothers for $10,000. Later, when the film was printed and shown, despite the inclusion of much of our work we were nowhere to be seen on the list of credits.

In the last couple of days there have been radio reports of more people converging on us than expected. One report says hundreds of thousands. Michael's amazed. He thought he'd get an audience of about twenty thousand, now it looks like it's going to be upwards of a quarter of a million. Warner Brothers are on the phone to him about film rights. The crowds are beginning to arrive.

After the show we rested up for a week. In the course of

conversation with Alan I'd mentioned an idea I'd had; Oedipus Rex as a Western. He liked it. He appreciated there were aspects of the Sophocles play that bore similarities to the historical and mythical West; recognition of powers beyond the control of Man, superstition, fortune telling, violence. He put us in a house on the West Hampton beach giving us three months to sort it out. It was my first attempt to write a screenplay and it did not come easily, it hardly came at all.

My efforts were interrupted by a phone call from Alan asking us to put it aside for the time being. He wanted us to make a film of Dr Timothy Leary, high priest of the LSD drug culture, running for Governor of California. He's got the budget. We're on the move again, a high profile, controversial subject to get creative with.

The three of us fly to California to discuss it with Tim. He was living in Hollywood at the time and we met him for a drink at the Chateau Marmont where we're staying. He loves the idea of the film, sees its campaign value immediately; gets quite carried away. He's not a filmmaker, of course. He sees filmmaking as an exchange of energy, an interactive social game. "We'll have fun." he says. "You film me a bit then I'll film you a bit." Mike and I exchange glances.

First day of filming we fly down from New York to Washington with him. It's a special day, a Viet Nam war moratorium, a good place and occasion for him to start his gubernatorial campaign. "Fuck the war! Turn on, tune in and drop out!"

Wavy Gravy, (name given to him by B.B. King at the Texas International Pop Festival in 1969), originally TV stand-up comedian Hugh Romney, has come up from the Hog Farm in New Mexico to lighten the proceedings. Washington cops, macho in riot gear, can get unnecessarily heavy with peaceniks. There's something about peaceniks that can really get up a cop's nose. Maybe it's their easy going ways, the long hair, sloppy, colourful, unconventional dress, who knows? Wavy's wearing a jumpsuit, a flying helmet, a large Donald Duck beak over his face and carrying a four-foot, inflated plastic banana. We film him recounting to the future Governor of California how he'd charged the police lines at Dupont Circle, a lone figure wielding the huge banana charging at the phalanx of heavily armed and armoured cops and how they had fired tear-gas canisters at him to drive him back.

Second day of filming Tim gets busted; I believe it was because of a joint discovered in his daughters clothing. Federal agents had been gunning for him for a long time on account of his druggy, anarchistic ways, looking for any excuse to take him out of circulation. Tim goes to jail. Another movie bites the dust.

Within days, Alan calls again. He has access to forty-two Laurel & Hardy sound movies, half the entire body of their work. He wants us to cut from them a composite feature-length film. He's talking to one of the Rolling Stones about a sound track. Warner are interested.

This is my very good friend Stanley Laurel and I'm Oliver Norville Hardy. The idea of cutting together a completely new plot out of all their nonsense appeals to me. Mike and I lay back and watch reel after nitrate reel of these two crazy geezers going through one routine after another.

It would be impossible, arrogant to even try to make another Laurel and Hardy movie in their image. No one could ever cut up their movies and reconstitute them as another L&H film with their kind of humour. No one could do that, not in a million years. I'm curious to see what happens if we cut out the mitigating double takes, those idiot gestures that turn potentially serious situations into comedy? What happens if we take out the laughs?

Alan likes the idea and arranges for us to have a cutting room we can use at night. I start looking for threads, clips I could play with that could be patched together to become themes. We try a few scenes and the effect is bizarre. Innocent scenarios turn into sinister plots. It's as if we're digging beneath the surface of these two adored idols, revealing a darker side. The Secret Lives Of Stanley Laurel And Oliver Norville Hardy.

We sleep by day and meet at the cutting room at eight in the evening, break for a meal at midnight down the street at the Stage Deli, then work through until eight in the morning. In a twenty-four hour city, the day shift gives way to the night shift. We like the night shift. Musicians, actors, filmmakers, prostitutes, gamblers, barkeepers, restaurateurs are all night shift people.

There's part of Mike that enjoys dealing, likes to bargain, likes to gain the edge. The night shift suits clandestine activities. He's buying small quantities of cocaine, cutting it and selling off a gram now and then. He's become a modest but regular user.

Although she seldom comes to the editing room, Barbara keeps the same hours as Mike. She likes wheeling and dealing as much as he does, demonstrating an affection even greater than his for their product.

Sometimes one or two of Alan's musicians drop by the editing room from all-night recording sessions. We take a break, smoke a little grass, snort a little coke. They're interested in what we're doing and follow the progress of the film day by day with anticipation like it's a serial.

I'm working with one of the recurrent themes in Laurel & Hardy, the escalation of violence. I'd seen Bruce Connor's film Cosmic Ray and been impressed by the way he'd built up a rhythm of disaster cut to a driving Ray Charles sound track. Girl drops plate, boy slips on banana peel, man falls off bike, increasingly complex bike and automobile pileups, train crashes, airplanes crashing as they land, bursting into flames, bomb and shell explosions working up in a frenzy to the ultimate catastrophic slow-motion explosion and mushrooming cloud of an atom bomb.

The scenes of L&H violence invariably start with an exchange of trivial mishaps and insults. Ollie parking the car accidentally bumps into the one in front breaking one of its taillights. The owner gets out, examines the damage, returns to his car, takes out the crank-handle, strolls back to Ollie's car cool as a cucumber and smashes one of its sidelights. Ollie, exasperated, examines the damage, walks to the front of the other car and smashes one of its headlights. This tentative tit for tat develops into serious property destruction and then into physical violence.

I find footage of Ollie on a soapbox, encouraged by Stan, vigorously addressing a crowd. Then I find a clip in which someone with a rifle takes aim and fires a shot. From another movie a clip of people running in panic across plazas, down streets and from another film, mass riots, then soldiers marching off to war, onto the battle field, charging, bayonets fixed, bombs and shells exploding, machine guns chattering, biplanes bombing and machine-gunning the trenches.

L&H walk away from it. They are the villains. While gallant lads are being mown down by machine guns in the mud of Ypres, they're boozing and whooping it up in nightclubs, two-stepping with beautiful, scantily clad girls. It's a collage of course, like *The Arousing*. No attention paid to continuity of set or costume, only to action and dialogue. It works.

The finale begins with a clip from a hospital scene. Stan unwittingly sits on the needle of a hypodermic syringe. Big close up. He's concerned; he complains to Ollie he's feeling strange. We splice together all the 'strange' material we can find in the films, ghosts, phantom ships in the mist; haunted faces appearing and disappearing, L&H rolling in slow motion down desert dunes, one scene dissolving slowly into the next, billowing clouds of steam over boiling water, an egg pops to the surface. The End.

I have that feeling again, the way I felt when we completed The Arousing. We've allowed it to happen and arrived at something extraordinarily special. We show it to our regular night visitors. They agree. It has coherence. The actors look familiar but their behaviour is not. Watching easily identifiable idols behaving uncharacteristically is a strangely humorous experience. It runs sixty minutes. A remarkable film.

Alan admires it but has his deal with Warner Brothers structured around a ninety-minute product and I've only produced sixty. He's cool. Encouragingly, he asks us to look at it again, see if we can't add another thirty minutes to what we've already done.

I've looked hard at the material, made it my own, made the only movie I can see in it and it only adds up to sixty minutes. Laurel and Hardy never made a movie longer than that but Alan wants ninety. I can't see one minute, let alone thirty, to add to what we've got. We've cut an extraordinary Laurel and Hardy movie but it'll never see the light of day because it's only sixty minutes long. It's insane. I've had enough. Mike's staying on to work with Alan, to pull the film apart and start again. I wish them luck. Alex and I are returning to England. No hard feelings. I'd produced another original, done something not done before, a fascinating metamorphosis of well-known well loved comedians from clowning to dark comedy. Making it was great fun. That it will never be seen is disappointing. It isn't the day for fame and fortune.

100.

Arriving back in London Alex and I crash with Bob and Kelly. The abortive attempts in America to complete a film are beginning to weigh on the mind, sapping the confidence. It's been two years since we left London for New York to bring enlightenment to the moving picture industry. Bob tries to get me to relax but I can't. I'm in a hurry, almost desperate to get something going. Making *The Arousing* was a consummate artistic experience. The Secret Lives too. I want that feeling of completion again but can't find it.

Before we went to Nepal I wrote and sold a story, *Pick A Card.. Any Card*. I'd been so taken up In America with present time and all that was going on, I'd forgotten it. Now it was the only vaguely promising card in the hand.

I call the producer who bought the story. He's delighted to hear from me. Where've I been? They've been trying to find me for months. They finished the movie, they call it *Vanishing Point*, and I'm just in time for the private screening. Suddenly everything is different. Success rides a steep curve; it takes your breath away.

When I create a movie in the mind I direct in the mind as I go. I can't help it. It's the process and I hope the person that gets to direct it will see it the way I see it. At the preview I watch the screen, watch my hard, steely story being invaded by characters and scenes I'd not invented, characters and scenes sentimentalised into acceptable Hollywood mulch. That's all I can see, the same old diminution of truth, the corny, the trite. Forget it.

Not that I'd held any hopes for *Pick A Card* but coming smack on top of the Laurel and Hardy fiasco it's all too much. I need a change, another shake of the dice. I'd like to dive into another film, another poem, something personal, but there's not enough money left to finance anything and my man, my partner Mike, is still in New York trying to cut ninety minutes of Laurel and Hardy.

101.

Writing's now the closest I can get to actually making a movie. It's all I can afford. There are a couple of ideas I'd like to explore. Alex and I will go away, find somewhere cheap, warm and quiet. Morocco. Neither of us has been there.

Alex is apprehensive about Arab culture, unemancipated womenfolk and all the other cultural differences we might encounter; even so we arrive in Tangiers with camera, typewriter and a considerable amount of baggage. Alex travels with everything she owns, an old habit hard to die.

A block away from its charming seaside facade of palm trees and white adobe, Tangiers is a pastiche of a grey concrete European city. We get burned in the first few hours trying to score something to smoke. A friendly American girl sells us henna as hashish. We lose no time getting out of the city. We travel by bus, east along the sun-smacked Mediterranean coast, past Tetouan and the pale blue village of Chechouen to a fishing village on the beach, Oued Laou. A café hotel, a few small houses belonging to local fishermen and, fifty meters from the sea, across the road from the café, a large, two storey, stone fort, the Spanish governor's house in colonial times. It's for rent.

It's massive, cheap and appeals to Alex's sense of style; it was the Governor's house after all. We move in. None of the locals care to live there; it's far too public. Provincial Moroccans are private folk. The windows of their houses face inwards onto courtyards. Outside walls are blank, expressionless with no indication of a family's identity or their status in the community.

The main room on the second floor of the fort is large with a panoramic view of the Mediterranean through its two tall windows. We live in this room for six contented months. Downstairs it's dank and creepy, lots of small dark rooms, spiders, rats, probably prison cells in past time. We don't spend any longer down there than we need, only to come in and go out..

We buy straw mats and blankets at the local *souk*, a couple of pots for cooking and a charcoal burner. We cook, eat and sleep in the room. We contemplate the changing colours of sky and sea framed by the white sash of the windows. Sometimes I take photographs by evening light, the sky grown pink over a pale, unbelievably peaceful, aquamarine sea. London and New York are far away.

I start writing. I'm smoking *kif* and thinking about memory, the vast library of information, learned or genetic, embedded in the brain. I arrive at the notion of receding memory, a pathological condition brought on, perhaps, by some disturbing experience. Receding memory is a gradual blanking out of the mind's old information. The mind is increasingly unable to hold on to new thoughts and wanders rudderless, closer and closer to the nick of time, ultimately foundering on the pin-point of the here and now.

It was good dope. In the course of buying it I make a friend. One of the few houses on the beach belongs to Salah, a fisherman. of indeterminate age who can generally be found sitting in the shade of his house mending fishing nets. On the side he's the local dealer of many things. When the fishermen's donkeys laden with freshly caught fish climb the Riff Mountains behind the village up to Kitama, they return with *kif*, a speciality of the region, for sale along the coast.

My visits to Salah are always pleasurable. We smoke and discuss the universe in faltering French for hours on end. Uneducated in any formal sense, he's a natural philosopher with an understanding of the world, its geography, it's history, its politics, its future. He understands the relationship between the infinitesimal and the infinite.

A cut sapling - I'm not sure of its specie - stretches across the main room of his tiny house, jammed between wainscot and ceiling. His father planted it when Salah was born. It was Salah's tree. When his father died he cut it down and stuck it in his house. It became his medicine tree. When anything ails him, he scrapes off a little bark and drinks its infusion.

We're staying here on the Mediterranean through the summer. When the weather cools and the rains come, we'll pack our bits and pieces, take a bus south and look for somewhere dryer and warmer.

We discover Taroudant, a sixteenth century town protected by five hundred year-old crenulated mud walls. We rent a recently built house in an olive grove a kilometre outside the walls, lay out our pots and pans, straw mats, blankets and settle in. It isn't any warmer. It's raining as hard as it is on the Mediterranean and the roof leaks.

We huddle in woollen *burnooses* around the primus stove brewing sweet mint tea. The drips from the ceiling tinkle into cans and tin cups spread around the floor. I take it up with El Hajj our landlord who speaks neither English nor French so couldn't understand why I was so insistent he comes to the house immediately. He must have thought there'd been an earthquake.

He laughs when we get there. When he can get a word out, he points to the ceiling, says between giggles in Arabic made understandable by gestures,

"It's nothing... it's nothing."

He was genuinely relieved.

"It's a new house. New houses always leak for a couple of years. Didn't I know that?"

I spend my time writing, taking a few snaps, shopping, cutting kif to smoke. Alex takes care of the cooking. We fall into a routine, relaxed and laid back. We hardly see anyone, no one that speaks English, no one to discuss the universe with. In town, I get along in French with one or two of the traders. Sometimes of a morning the local Police Chief comes around and we sit outside in the sun and smoke a few pipes. He speaks a little French and we exchange pleasantries. Beyond that we're on our own all the time and when Alex's the only person in the world to talk to, it doesn't add up to much.

It's Ramadan. We're on the move, breaking out for a while, travelling by bus all over Morocco. We don't think of ourselves as tourists. We live here but don't know much about the country beyond the immediate vicinity. Some people, before visiting another country read up about it, want to know everything about the place before they get there, where to go, what to see, where to stay, what to eat, what not to eat, how much to pay. It's understandable if you've only got two weeks. It's different when you've got all the time in the world and enjoy surprises.

Were spending the night in a French-style *auberge* high in the Atlas Mountains. Other than a young Canadian couple on their honeymoon, we're the only guests. We've been living in a non English-speaking community for a few months. Meeting a couple of Anglophones I find myself moving in on them. Also starved of English conversation, they are as keen as I am to chat. We talk about our lives and what we do. We're talking movies and as I'm telling them about the making of *The Arousing* their brows furrow, eyes narrow like they're trying to remember something.

"Wait a minute... We've seen that..."

On their wedding night in Toronto they went downtown to see a skin flick. *The Arousing* was the warm-up short. They recognise Alex.

Now we're in Casablanca. It's not the romantic desert oasis with palm trees built on a studio lot in Burbank, it's grey and formal like a nineteenth century Paris suburb. We have an address of someone in Casablanca. When you're leaving home for foreign parts, everyone's got someone for you to look up when you get there. We have the address and phone number in Casablanca of a Dutchman, Joss de Blank. We telephone him. When he answers I can hear strange, unfamiliar music loud in the background. I explain the circumstances. He's delighted to hear from us and, like we're old friends, invites us over.

The brass nameplate outside the building says he's a dentist. We climb the stairs to the top floor and follow the corridor to the source of the music. I ring the bell. The door opens letting out a cloud of hashish smoke. Joss, in white dentist coat, greets us smiling. We follow him in.

Weird, antiphonal *Gnawa* music is loud on the stereo, shrieking pipes and incessant drumming. Writhing to it with a tall Moroccan youth, an attractive twenty year-old European redhead. Immersed in their dance, they take no notice of us,

A young guy, blond hair like a lion's mane, reclines in the dentist chair smoking a joint watching them. We sit and watch them. He turns to us, passes the joint and introduces himself as Tim. He's friendly, has an American accent. The music ends,. Tim introduces us to the girl as Cheetah. He didn't seem to mind the intimate way she danced with the young Moroccan. Tim has hashish and cocaine. Joss wheels in a big tank of nitrous oxide. We party with them for the rest of the day.

102.

In the spring, Alex and I return to London. We've been away a year and I'd completed an outline of the memory story. It now had plot and characters. I like to think it's a reasonable piece of work, original, intellectually sound but I'm not crazy about it.

Still without a home of our own, we are crashing again with Bob and Kelly. We visit Barry and Jacqui. Barry's publishing business is doing well. Jacqui goes to the press a couple of days a week to do the accounts. She's enjoying it, happy that at last there's a little money coming in. Barry's latest publication Contemporary American Verse is highly praised by academia and publishers alike. She's right proud, puts her arm around him, chatters on about it. Barry studies his fingernails, bashful when someone sings his praises. Nothing seems to have changed. Inside the family we've become, everything is warm, positive, encouraging.

I present my story to a few producers I know. There's no instant response as there'd been with *Pick A Card*. They all remark on its excellence, its originality, producers hate to hurt. They all say how well it's written but it's not for them. I didn't have much riding on it; I'm not bitterly disappointed. But money is running out and I need to earn. I need to get involved one way or another with a movie.

Kelly's not well. She never complains but you can see her discomfort, curled up in an armchair, trying to smile, denying anything's wrong. You never know what's going on in another mind. She's a shy thing. Out on the catwalk she's a confident, dominant beauty, but down here on earth, underneath it all there's a timid brown mouse of a girl.

I'm languishing in these doldrums when Candice unexpectedly

calls from Paris. She'd like to see me for old times sake. "Would I like to see her?" Would I like to see her? I'm already imagining a return to old times. I'm being drawn back hormonally into what was a hopeless relationship, all the downside forgotten, the fights, the pain, the totally unacceptable behaviour, all gone. Only the hormones remember. Would my hormones and I like to see her? Of course we would.

Alex hasn't actually met Candice but well remembers that night in New York, the phone call that shot her so precipitously out of my life. She knows enough about her and aware deep down that I've never really gotten over her. Now Candice is hovering about in her life again, as threatening as ever. I don't try to explain. What can I say? I don't really love you Alex. I still feel passion for an old lover who's a pain in the arse?

She's in tears; says she's had as much as she can stand of my bullshit. She's right. I'm being an arsehole. I know all this but do nothing about it, nothing to reassure her. I feel the last vestiges of resolve going down the drain, only one thing on my mind.

Alex packs her belongings and moves to her stepmother's, a tearful parting, sad and tearful. I'm guilty as all hell but counting the minutes.

Candice arrived looking more beautiful than I remembered. That faint smile at the corners of her mouth, just the touch of her hand sends shivers of delight and desire chasing through me. She knows she still has the power. Bob and Kelly are pleased to see her again. It's like old times.

Next evening I show her off to Barry and Jacqui. She fascinates them. They make a fuss of her. Candice enjoys their attention and by the end of the evening we're all making love together, not exactly what I'd had in mind. The following morning when it's time to leave she says she's staying. She gives me that look half way between amusement and pity, raises her eyebrows and shrugs.

I've lost the bitch Candice. I'd lost the gentle company and friendship of Alex. It took a while for feelings of disappointment and trashed ego to subside. I'm roused out of this stupor of self-pity by a phone call from Mike. He and Barbara are in New Mexico getting a movie together. There's money from a local drugs smuggler who wants to spend his profits on a movie about his search for the best marijuana. *The Quest For The Best.* Mike wants me to come and direct.

A movie about marijuana? Funded by a dope smuggler? Santa Fé? A long shot if ever I heard one but I'll consider anything that'll get me out of where I am at the moment. I tell Mike it's a great idea. Why not? I'd always wanted to visit Santa Fé. I'd heard Billy the Kid's trigger finger lies pickled in the museum there. Mike sends me a ticket.

I'm on the plane. The past is where it belongs. No room in the mind for Candice, the disappointment, my jealousy of Barry and Jacqui. No room for embarrassment at the way I treated Alex. I'm on the road again; off to make a movie and the mind is full of possibilities, images, soundtracks.

I have a 24-hour stopover in New York. I call Ed from the airport. Ed's a friend of Bob, a really light-hearted guy, easy to be with. He said I should look him up when I was in New York. He has little more than a walk-on part in this story but he's a catalyst to important events. He'd just come from sailing a boatload of Colombian marijuana across the Atlantic to Hamburg. He tells me after they'd made their delivery, they were in their Hamburg hotel room counting the money when three men, one with a gun, burst in on them, money or your life. Ed said he grabbed the only weapon to hand, an acoustic guitar, sprang onto the table money flying all over the place with the guitar raised above his head, and screamed at them to get the fuck out. Surprised, they looked at him like he was crazy and backed out of the room

It's 4 a.m. Ed and I are in a Greenwich Village recording studio listening to a group, friends of his, recording one of their songs, The Ballad Of CC Younger and Del Rio Dan. I listen. The lyric is about two young Texans backpacking marijuana across the Rio Grande.

The movie's beginning to make itself. At the end of the session I chat with the songwriter. He says CC Younger and Del Rio Dan are real people; good friends of his and constantly in touch. I give him Mike's address in Santa Fé. I want them in the movie. I've got the music and my actors.

It's evening when the plane touches down at Albuquerque airport. Mike and Barbara are there to meet me and it's good to see them again. We light up a spliff and drive out of Albuquerque into the darkening countryside.

The Laurel and Hardy movie didn't work out. I could have told him. Mike's a good cameraman and under direction a good cutter. But he doesn't have the mind for story telling, doesn't have that kind of imagination. We talk about the new movie, *The Quest For The Best*. They've roped in an actor to play a lead role, a minor Hollywood star that happens to live around Santa Fé. He's not needed. I've already got my actors. I tell them about the song and the young Texans. They're excited. We're definitely on our way. I'm back on track. Life's turned into a movie again and I'm feeling good. Mike says there's a meeting tonight with some of the people involved, a kind of script conference. We're going there now. Fine. I'm keen to get started.

Dark shapes of mountains loom against the deepening purple sky. We're somewhere in the Jemez Mountains driving up narrow dirt roads. We arrive at a ramshackle brick building difficult to figure out in the dark.

Mike leads us through doorways and passages into a lighted area, an old concrete swimming pool of murky steaming water, a hot sulphur spring.

At one end, under dim yellow lights, half a dozen naked men and women relax in the warm water. They're not discussing a script; they're not discussing anything. They seem in contemplative mood, almost catatonic, moving only when it's their turn to take the joint passing between them.

Mike, Barbara and I take off our clothes, Mike unbuckles his leg. We slip into the warm water with them. No one makes any introductions. Who cares? I'm too stoned to give it much consideration. A powder is being passed around with a straw, something new called *Angel Dust*.

Morning. I wake up in a house in Santa Fé, a large adobe house belonging to Igor, a successful New York society portrait painter, friend and client of Mike and Barbara, a pleasant, civilised man. He knows Candice. Knew her in Paris, says she left a hickey on his heart. After she and I separated in Paris she fell ill, ending up in hospital. That was one of her ways of resolving dead-ends in her life. She'd get sick, move into a hospital where she'd wait to see who's going to come along to love her and save her. Igor, living in Paris at the time, was there every day, bringing her flowers and chocolates, caring for her with devotion, sleeping at the foot of her bed every night.

There are others at Igor's house, mostly from Manhattan, arriving, staying a few days, leaving to be replaced by new arrivals, an ongoing house-party. Mike and Barbara supply the euphorics. Mike has taken to wearing a tiny silver spoon on a thin chain around his neck, a birthday present from Barbara. He seems to be spending more time dealing drugs than making films.

Barry shows up unexpectedly. I hardly recognised him. He'd

shaved off his beard and moustache, grown his hair long to his shoulders. He's lean and film star good-looking, a transformation from the studious printer-publisher in West Hampstead going unobtrusively about his business. He's in sombre mood.

After I'd left them in London, he, Jacqui and the children and Candice went to Minorca as a *ménage*, staying with one of Candice's wealthy connections. He and Candice fell in love. When they returned to London he tells Jacqui his feelings. He needs some time with Candice, just a month to work it out, he'll be back.

He didn't work it out and didn't go back. He's just come from New York where he'd been living with Candice in her midtown apartment. It didn't last long. They had a bust-up over money. He didn't have a penny; she was earning big and started treating him like a poor relative. Told him to get off his arse and earn his own. She's a mean motherfucker, the meanest he's met. She's history, never wants to set eyes on her again.

There is an understanding between Barry and I about Candice as though we'd campaigned in the same war. We'd both loved the same woman, been fucked over by her and that's all there was to it. I ask him about Jacqui and the kids? He mumbles something about going back to London to take care of business. He'll talk to her then. He's pretty miserable about it. Despite outward appearances, he hasn't really changed. He's still Barry, taciturn, forever preoccupied with his own thoughts. But there's an edge to him I hadn't seen before. I'd only known him as Jacqui's partner. He's different on his own, more assertive, a little more outspoken.

Today there was another *Quest For The Best* script conference. Mike assured me everyone was serious, no more sulphur-baths and angel dust, we're really going to do the business. The producers had borrowed Dennis Hopper's house in Taos, a fine old adobe mansion that used to belong to Mabel Lucy Dodge, where her lover DH Lawrence had lived with her. We drive up there; Mike and I find our way through the house to the conference.

Spread around a table, six or seven longhaired cowboys smoking joints. I think I'd seen some of them in the sulphur-bath but couldn't be sure. At the head of the table, armadillo-booted feet on the table, a confident-looking guy I'd not seen before chairing the meeting. He refers to himself as the producer and director. I have a bad feeling.

Mike whispers I shouldn't take any notice of him; he's just a bit-part Hollywood actor. Maybe he is, but right now he's acting the producer and director on a film I've travelled a long way to direct. He's waffling, Hollywood style, He's faking it. It's obvious he hasn't a clue how to make a movie.

This isn't a movie; it's a stage for this arsehole to parade on. He makes it clear he's running the show. The rest of the men around the table either know as little about filmmaking as he does or they're too stoned to care. There's no room in it for me. What can I do? Call him on it? Have a shoot-out? I could have tried to present in a serious way the ideas already formulating in my head but the atmosphere riled me. I just wanted to get out

Mike and I drive back to Santa Fé. I'm furious. This is the fourth time a movie has slipped away from me. Sure, these kinds of projects are unlikely to come to anything but you have to follow every lead. Opportunities offering a budget, actors, a theme, *The Ballad Of CC Younger and Del Rio Dan* are not an every day occurrence. But as it turns out, it's not an opportunity. I thought I'd got it right this time but I was wrong. So it goes.

CC Younger and Del Rio Dan materialise for a few hours, they show up at the house in an old camper truck. They're everything their friend in New York sang about, the real thing. A couple of carefree young Texan backpackers who make a thousand dollars a trip wading across the Rio Grande with a couple of kilos, sometimes dodging Federale patrols as they scramble into their truck, bullets pinging and whining around them; they show me the bullet holes. They sell on the marijuana and then go fishing. And here they are, with fishing poles, asking when we start shooting the movie.

I explain. They're cool. You have to be philosophical, swallow disappointment and get on with the next. Dan rolls up a fat one and we smoke. You can't read the future, you never know what to expect. I feel lighter; glad I met them. They spend the night with us and leave early in the morning.

103.

Barry's shopping in a liquor store in town and gets hit on by a gay actor who invites him up to the set where he's working in a movie. It's in the mountains outside a little township in southern Colorado called Pagosa Springs a hundred miles north of Santa Fé. The movie's a western starring John Wayne. Barry can't resist and drives up there; he's a big fan of the Duke. The gay actor is not on the set the day he arrives. That's OK but neither is the Duke. He hangs around the set. Two attractive young female extras catch his eye and when the day's shoot wraps they take him home with them.

Rachel and Tina live on a ranch, a massive spread the other side of Pagosa. It's not theirs. They and their man Jim are the caretakers. Their man Jim? He's away and won't be back for a week. Barry moves in and they party. Back in Santa Fé, he says we should both go up there. I'm at a loose end.

The ranch is about three miles outside the little town amid rolling green pasture-land bounded by forest, all sheltering under the dark, snow-capped Rockies. Breath-taking beauty. We enter the house and find our way to the kitchen. Jim's back. Wearing dark glasses, he sits at the kitchen table, Rachel standing one side, Tina the other. They're brushing out his long dark hair. On the table in front of him lies a disassembled .45 Colt automatic. He kept it when he discharged himself from the marines as psychologically unsuitable. He's cleaning it, the girls are brushing his hair, the camera of the mind is turning over. They're pleased to see us.

Jim's a cowboy hipster. He deals drugs and Navajo and Zuni jewellery. In the hunting season, he packs out for city-slick hunters on the mountain. City slickers sometimes make a kill but don't know what to do with it. They don't mind killing the animal, they're right proud of that and can already see the antlers on the wall. But they don't want to touch the dead beast, don't know how to get it home so they can show everybody. Jim disembowels the kill for them, ropes the carcass on a horse and brings it down the mountain. Nothing to it for a real cowboy.

He's not really a cowboy. He came up here when he stopped being a marine, attracted by the style, the macho ethos of the mythical old West. He comes from the San Fernando Valley in California and I guess he played Cowboys and Indians in the valley when he was a kid like Barry and I did on London Commons. Being close geographically and historically to where it all happened, it wasn't difficult for him to continue the fantasy into his adult life. There are still broncos to bust, steers to be herded and driven and it's legal to bear arms and spin yarns.

The girls are also refugees from the San Fernando Valley.

Everybody we meet seems to have migrated here from Southern California, most of them searching for some kind of meaningful existence, a life away from the noise, the smog and the California soap opera they were born into.

Jim admires my brand new, pale-grey Stetson, picks it up and looks at it from all sides, says his father used to shape his Stetson the same way. A Stetson doesn't get the right feel to it, doesn't look right, he says, until it's had the treatment, something he learned from his dad. He opens a bottle of Jack Daniels and pours some over the indentations in the Stetson that give it shape and style. He presses them in tight with forefinger and thumb as if setting them firm for life. He hands it back to me ceremoniously like it's now a crown.

I stare at it ruefully. It takes me a few moments to appreciate the effect. It used to look like the kind of Stetson you see on dude cowboys on fancy vacation ranches, shamefully clean and well brushed. Now, stained by the whisky, it looks as though it's seen a thing or two, been places, soaked up a little sweat. Not so bad. I can live with it. I thank him.

We stay up there for about a week and through Jim meet all kinds of mountain folk. Silver Hand, another one from the Valley, used to be Joseph, a ladies' hairdresser, now he's Silver Hand. He makes exquisite silver and turquoise jewellery better than any Indian. His delectable partner, Running Stream, formerly Darlene Rosebaum, makes finely worked traditional garments of Elk skin. We're in their company quite often. I take a shine to Running Stream and she shows some reciprocal interest.

The next time Barry and I call on them, Silver Hand opens the door to us, a levelled pistol in his hand. Looking me in the eye, speaking quietly, he tells me to stay the fuck away from his woman and closes the door on us. He couldn't have been clearer than that. On the drive back to Santa Fé, Barry muses about bringing Jacqui and the kids there to live. It could be a great life for them, certainly an improvement on West Hampstead.

At the house there were a bunch of messages. Candice has been calling Barry from New York. She wants to talk with him. A few more calls back and forth and Barry's packing his bag. He grins sheepishly when we're saying our farewells. I know what he's going through. All reservations, all uncertainties evaporate to nothing when you're in love with Candice. The impossible past is a blank, forgotten like it never happened.

Mike's still involved with *Quest* but getting bored with the whole production. They've taken all this time shooting one sequence, less than a minute, the Hollywood mini-star as cop stopping a car reflected in his mirrored sunglasses. They snorted the rest of the budget and are asking the dope smuggler for more money. Mike doesn't really care. Film's no longer his only option and already taking a back seat. He's talking about walking cocaine out of Columbia, says the money's fantastic.

I'm packing my bag. The movie didn't happen; my partner's become a coke dealer soon to be an international smuggler. Things didn't pan out the way they might have. I wonder about this movie career that started with a roll of drums and clash of cymbals. It's not enough to be talented, to have original ideas. You've got to stand in line jockeying with a bunch of egotistical cowboys seeking the bubble-gum Hollywood reputation. Right now I want to be in more familiar surroundings. I want to curl up with friends on an English hearth. That's how I'm feeling.

The spare bedroom in Bob and Kelly's Kensington flat has become my regular *piéd á terre* in London. Bob meets me at the airport. He puts his arm around my shoulder as we walk to the car; he can see I need some loving. He lights a spliff and passes it to me as we drive off towards the city.

Being with Bob is easy. There was always mutual respect from the first time we met. We don't say much when we're together these days. We're not gossips and don't do much philosophising any more. We used to talk a lot about words like god and truth but now we're just good company. I tell him about the film fiasco and how fucking depressed I feel. He laughs; I laugh. There's a funny side to everything. We're in Kensington, parking in front of the flat, before he breaks his news. He tells me Kelly's condition has worsened. She's got cancer. That's it. Nothing funny here. No funny side I can think of.

She doesn't look good. Her warm brown skin has a greyish hue. Eyes dark-circled, she smiles up at me from her armchair as though everything is just fine. I kiss her and ask her how she's doing. She shrugs, makes a wry face. Says she's got cancer. What can I say? What can anyone say?

She smiles, chats about the modelling business, how well she's been doing until this. Ossie and Alice can't wait for her to get better; no one can show their clothes like she can. All her friends are supportive. You can tell she's trying to get a grip on things, trying to hold off the spectre, trying to keep it out of her thoughts. Lovely Kelly is dying as life goes on around her.

While she's consciously awake, Bob is the epitome of calm, gentle with her, attentive to all her needs, which are few. She spends most of her time daydreaming, in and out of sleep. While she sleeps, he's on the phone wheeling and dealing to keep his mind focused, putting something together, anything.

Al Vandenberg called, the art director from DD&B, New York, the one I didn't much like who introduced me to Bob. I'd forgotten about him. Bob told him I was on my way to London and he wants to meet. How about the Picasso in the Kings Road?

Even though I owe him for my friendship with Bob, I'm not looking forward to seeing him again. I'm not in the mood for his high-speed, New York bullshit. But four years in sixties London has wrought changes. I hardly recognise him. No more suit, no more button-down collar. He has long hair, a beard, jeans, sneakers, T-shirt under a dodgy looking military-style jacket and his faithful, one-lens Leica.

He's been having a great time. Works for fashion magazines; pulls out slides of yesterday's shoot. They're excellent, perfect colour, perfect focus. Beautiful young girls looking out at him like he's an old friend. Al's never had it so good and getting paid for it. He invites me back to his flat for a smoke. It's just down the street from my old Chelsea studio.

The fireplace dominates his front room and rising from the mantelshelf and spreading across the wall behind it, a monumental collage of *images trouvées* clipped from newspapers and magazines mixed in with some of his own snaps. Threads of Buddhism weave through a colourful confusion of graphics and pictures of people. On the mantelshelf, everyday shit turned into holy shrines, sacred oriental soap and cigarette packages, matchboxes, an altar to some latter-day god of twentieth century graphic art.

He's rolling a spliff. We smoke. He talks, I listen. He's having the time of his life. Plenty of sex; working with *The Beatles* on a new album cover; friends with important fashion editors and fashion designers. He's taken a lot of LSD and smoked a lot of dope. I can believe it. The Madison Avenue veneer has completely gone and we're beginning to see each other.

He tells me about himself, where he comes from and so on. He doesn't know his mum and dad, an orphan from the outset. From the orphanage into the army, two tours of duty in Korea; wounded a couple of times. Then art school; advertising; marriage; separation.

He left his ten-year-old son with his ex wife. Now the boy's fourteen and causing havoc in Philadelphia where he lives with his mum. Truant from school, in trouble with the police, she's constantly writing to Al about it. He feels guilty. Thinks he should have done better. We shake our heads over Kelly's dying. There's nothing to say. We part better friends than we'd been in New York.

I called Alex suggesting we meet. Surprisingly she's happy to see me again and in no time at all moves in. It's not a reunion of old lovers, more like a meeting of estranged friends. When she nestles innocent under my arm, head on my chest, she thinks I'm her father. She thinks she's in love with me but she's really looking for a father. Whatever, I'm happy to be nestling up to her again, familiar smell, quiet, no surprises. At the same time some part of me remains constantly awake to the possibility of someone new and exciting entering my life, someone that will surprise me, an unpredictable Diva I can play with. Alex's started dressing like Candice, even adopting some of her mannerisms.

The telephone rings. It's Barry. He's in Santa Fé with Candice. He's sold his poetry business and its imprint to a major London publisher for a lot of money and bought a house. He has an idea for a movie, a documentary. He wants me to come over and stay with them, we'll do it together, share the profits. He'll produce, I'll direct. I'm awake again. A movie.

Alex and I visit Al to say good-bye. The street door's open. Alex smells burning. We find Al in the front room, comfortable on the sofa. The montage on the wall has his rapt attention. It's on fire, flames creeping to the ceiling. A candle on top of a plastic replica of a human skull, had burned down, set the skull ablaze and spread across the whole mantelshelf and the montage behind it. He's watching its progress with interest, too stoned to move. We run around extinguishing the flames. Only then does he notice us.

Alex makes coffee and we straighten him out. Hunched on the sofa, he lights a cigarette, sips at his coffee surveying the scorched wall, the dying embers, says God's been reading his mail. He's had enough of London and fast-lane fashion photography. He's packing up and leaving. He's going back to the States to look after his son. He'll take him to California out of the way of his mother. It's been on his mind. The kid will be OK there. Yeah. California. He's sure of it.

I tell him about the movie Barry wants to make. We're going to be near neighbours. It's the first time I give him a hug and feel his body try to respond. He follows us to the door. If we get to California we must look for him. Bob will know where he is.

Barry meets us at Albuquerque airport in a spanking new Chevrolet pickup. He's tanned, longhaired and handsome, elegant in a new Stetson and expensive armadillo boots. Lean bronzed arms adorned with heavy silver and turquoise bracelets, silver and turquoise rings on one or two fingers. A twelve-gauge shotgun hangs in the back window behind the seat, a gift. he says, from the bank when he opened an account with them. Barry was a chrysalis last time I saw him, now he's emerged full-bodied, wings spread. I'd seen it years ago. It was there all along waiting to come out. It took Candice to release it.

A few weeks ago he's in San Bernardino, California, driving down

Main Street. He sees a cluster of people around a used car sale. They're watching a demonstration, some kind of entertainment. Barry pulls over. The owner of the car lot had hired a fast-gun as a crowd-puller for his weekend sale, and there he was, on stage with his blonde, fast-gun wife, drawing at the speed of light and shooting the shit out of polystyrene cups and balloons. Pure Americana. Barry's fascinated. Rob Mundon is in the Guinness Book of Records as the fastest gun in the world, thirteen-one-hundredths-of-a-second to clear leather and hit a target. That's fast. We're going to film him.

It's dark when we pull into the driveway of a large, sprawling, brick-built, bungalow on the outskirts of Santa Fé. The shingle hanging by the front door reads *Confusion Cottage*. Barry says it was there when he moved in. He grins says Candice really liked it.

Candice's already gone to bed. Barry shows us to our room. Alex beat from the journey turns in. Barry and I open a bottle of tequila and catch up on history.

He and Candice had gone back to London, a flying visit long enough to complete the sale of his company. They'd gone over to see Jacqui but she wouldn't let Candice past the door. He could come in but she could go fuck herself. He grins, shakes his head at the memory of it. He drags out some guns he's bought, a matching carbine and pistol both with blued, octagonal barrels, 3030 calibre, collectors' pieces. We cock them and click them. He points out some nicks in the pistol grip. Someone had been keeping score. He says we can go to the city dump tomorrow if I like and blow the shit out of old cars.

I ask him about Mike. We're going to need a cameraman. Barry's uncomfortable about it, says he doesn't think Mike's right for the job. Since *Quest* fell apart a year ago Mike's been coasting, He's sadly out of practice, more interested in cocaine than cameras. Barry refills our glasses. This is a serious flick, his own money being spent. He's met a cameraman fresh from a shoot in New York. Seems a nice guy. He'd been working for Andy Warhol. He can cut, too. Can't do better than that, can we? What the hell, it's Barry's money and he's made up his mind. No point in arguing. I'd have liked to be working with Mike again, the old team. But it's true. These days his mind is on other things. We finish the tequila and I go to bed.

It's ten in the morning. I'm the only one up. The air is warm. I slept well and I'm ready to get on with life. I investigate the garden, a fenced-in half-acre of desert, rocks, sand and a few cactuses. The swimming pool's empty and in need of repair. I'm in the kitchen making coffee. Through the window, silent sun-baked hills and arroyos roll out to the distant mountains under azure sky, a tranquillity suddenly shattered by a loud, electronic buzzer, a timing device on the cooking stove's complicated panel of dials and knobs. The buzzing fills the head so early in the day, aggressive, persistent.

Before I can figure out how to turn it off, Candice, barely awake, wanders in, one hand holding together her robe, a large hammer in the other. She walks like a sleepwalker to the stove. I step aside. She kills the panel with fierce blows of the hammer, smashes it until the noise stops, then goes back to bed. She doesn't look at me, doesn't say a word. Comes in, hammers the stove as though it was the correct way according to the instruction manual to turn the noise off, then leaves without so much as a nod or a wink for her old lover.

I'd forgotten how it is with Candice. The mood at Confusion Cottage is her mood; unmistakable from the moment she gets out of bed. She can make the day carefree and light, all old friends together. She can be moody, vindictive and make it dark. She can make you feel like her trusted confidant or like a despised enemy. She'll draw you to her bosom, the next minute she'll spit you out in two-franc pieces, want you out of her house, out of her life. I'd forgotten.

She's co-producer. Barry's keen for her to be involved. It'll give her something to think about, keep her from getting bored. She wants us to know she wasn't without filmmaking experience. She'd brags about this lover, a truly hot Hollywood writer. She talks about Gene Grozotski the cameraman from New York as if he's her protégé, the new Lazlo Kovaks. Says it's not everyday a Warhol cameraman falls into your lap. I look across at Barry. He's examining his fingernails.

Gene comes over to the house to meet and discuss things. He's tall and well-built, short brown hair flopped across his forehead, handsome in a boyish way. He smiles, he's charming, breathes out greetings in a velvety New York voice. He's not at all what I expected coming from Andy Warhol. Candice cuts a wages deal with him that includes board. He's crew. He'll come and live with us. There's plenty of room.

Barry telephones Rob Mundon the fast gun and fixes a date, time and place to meet. It's another used car sale this time in San Diego. We'll watch his show and talk afterwards, Barry, Gene and me.

Alex is apprehensive being left alone with the woman who a few years ago threatened to kill her. She's already loomed large in her life but this is the first time she's keeping close company with her and doesn't yet know how to read her moods, how seriously to take her. She doesn't say anything but I see her apprehension. She says while I'm away she'll contact Mike and Barbara, maybe spend some time with them.

The three of us are in the Chevy on the outskirts of San Diego looking for and finding the car lot. We pull over. Not a big crowd, twenty or thirty. Rob Mundon's already on stage with his blonde wife, Doris. Both of them are short and well built, both dressed in khaki shirts and trousers, military-style Stars and Stripes shoulder flashes, cowboy boots, hard straw Stetsons and tooled leather gun belts.

They address the small crowd like seasoned professionals. Mundon strides up and down the small stage with a hand-mike, dramatising the act he's about to perform, making his audience realise how difficult it is, how smart he's got to be, how fast he's going to have to draw. Doris is setting up music stands with balloons dangling.

They go through their routine, drawing and blasting away. They use blanks, the cartridge wadding having sufficient velocity to burst the balloons. Mundon strides the stage, telling his audience about the mythical gunslingers of the old west, heroes and villains, how they wore their firearms, who fought who and how they drew.

At first I can't see beyond this plastic entertainment until I realise I'm watching a slice of pure Americana. Only in America. The great heroic tales of the West reduced to a rehearsed patter, the art of self-defence to a circus act. The whole thing looks a bit unexciting for a movie you might say, particularly if you're an American. But that's OK. We're not making a tourist commercial or a soap opera. This is a documentary of real life in the US of A. Man and woman, parents of children, law abiding citizens earning their living.

There are two kinds of revolver, double action and single action. Double action doesn't need to be cocked; you pull the trigger and it automatically cocks and fires in one. The pistol used in fast draw, is single action. The hammer has to be manually cocked. This means drawing the revolver from its holster with the trigger already pulled. The hammer remains free, the other gloved hand fanning across the hammer fires the bullet.

This is the finale. Mundon pulls a leather glove onto his fanning hand and places a polystyrene cup on the back of it. He's poised to draw. We're all waiting. He draws at the speed of light, fanning the gun into action. The cup, left momentarily unsupported in mid air, is blown apart before it has time to fall. I'm impressed. Even the crowd's impressed. A patter of applause.

We go to the Mundon trailer and introduce ourselves. He's cleaning his silver pistol. He turns, looks us up and down, from long hair to armadillo boots, finishes polishing the pistol and holsters it. We tell him we want to film him.

"If you think you can catch my draw at twenty four frames a second forget it. I'm faster than that. At twenty-four frames you won't see it on any single one of 'em. I draw and fire in thirteen-one-hundredths-of-a-second boy. It's in the Guinness Book of Records."

He smiles at us, eyebrows raised as if to say what do you think of that. Even though we already knew, we make signs we're impressed. We assure him we're going to film his draw in painfully slow motion.

He and Doris are amiable folk. They read through Barry's prepared contract. Barry knows about contracts from his publishing days, the need to get everything sorted beforehand, everything agreed, everything official. Something I'd never bothered about. The Mundons sign. Barry signs. We fix a date when we'll meet in the San Bernardino Mountains, a spot where the Mundons regularly work out. We drive off North up the coast to Hollywood.

We're in a facilities company on Hollywood Boulevard, searching for the slowest of slow motion cameras. I don't yet have a clue the way the film will evolve, what will happen in front of camera once we get going.

But I know that Mundon's thirteen-one-hundredths-of-a-second draw to shoot the cup is definitive, the most powerful visual event in his story.

The store assistant shows us a small white enamelled box, a lens at one end. It doesn't look much but it's a slow-motion camera, painfully slow, a thousand frames a second. He says Antonioni used it in *Zabriski Point*. Remember those slow-motion explosions? The exploding refrigerator? The cold chicken, the milk, eggs, orange juice, bits of salad, all exploding slowly outward, floating to the periphery of the frame and beyond. Boy! That was slow. We add the camera to the rest of the equipment we're hiring. Barry writes a cheque.

When he and I are driving together, as it is with Bob, we don't do much talking. We smoke, listen to Country and Western on the radio but don't talk much. True, we might exchange a thought or two from time to time but in the main we let the business of the day take care of itself. With Gene in the cab it's different. Gene's an affable talker, likes to know where you come from and who you know. He'll give you a not very original opinion on all sorts of stuff whether you want to hear it or no.

We meet up with the Mundons as we'd arranged and follow their camper into the mountains. It's a warm, sun-splashed morning, the air still cool and clear. We arrive at the entrance to a valley surrounded by mountains, the San Bernardinos, highest range in California. No water and little shade. Not a place to get lost in. We park.

Gene unpacks the camera. Mundon and his wife wrap gun belts around their waists. They don't wear them low on the hip tied with thongs to the thigh like gunslingers in the movies; they wear them tucked high on the waist. They don't have to reach down for the gun and raise it to the target. All it takes is a quick wrist action to twist the pistol free of its holster already raised.

We start filming. I'm running the Nagra for synch sound, Barry is clapper-boy. The Mundons draw against each other, Rob keeping up a running dialogue with his wife.

"Doris you're Buck Rogers and I'm Hoot Gibson. When I say draw we draw OK?"

Doris shuffles around in preparation waiting for Rob's signal.

"Draw!!!"

Blam! Bob's holstering his pistol before Doris even has a hand to hers.

"Gee Bob I wasn't ready."

"Doris. I keep telling ya. Stay awake girl! Gotta be ready at all times. Hoot Gibson is always ready. Draw!!!"

Again he drills her full of holes before she can draw her gun. And so it goes on throughout the morning until Gene's satisfied he has the scene in the can.

I have to take the Mundons game as seriously as they do or I won't get a film. It's amusing in a childlike sort of way, the dialogue, Mundon's silly banter, Doris pleading with him to play fair, not to shoot her full of holes before she has a chance to draw. Grown people, parents of children. I want to see the domestic side of their life, how they live with their two small daughters in a motorized camper. We agree to meet the following day at the trailer park where they live.

A bright young morning. Barry and I are eating breakfast in the motel restaurant. Gene's up later than usual and when he shows he's not in a good mood, nervous about something. He doesn't want anything to eat, just coffee. He asks to borrow the pickup, says he's got to go downtown.

Barry tosses him the keys. I remind him we're scheduled to shoot the Mundons this morning. He doesn't answer. He's gone. I ask Barry what's going on?

"Gone to see if he can score."

"Score What?"

"Skag."

I look at him. Disbelief.

"He's a junky?"

Barry is examining his fingernails.

"Yeah. Don't worry. It's OK."

Don't worry it's OK. The guy's a junky. Our professional cameraman is a fucking junky. It's midday and he's not yet back. Mundon calls. Wants to know where the hell we are. Barry tells him we've had a problem with a camera and Gene's taken it into town to see if it can be fixed. We'll be over as soon as he returns.

We sit outside the motel in the warm sun, waiting. We smoke a joint. Gene returns. He parks the truck and strolls over to us all smiles. He tosses the keys to Barry. He smiles down at us. Asks charmingly,

"Shall we go?"

Mundon is alone. Doris has gone to collect the children from School. He's pissed at us. He leads a very structured life. In his mind, time and space are a structure. There's a time for this and a place for that. You have to be structured with four of you living, eating, sleeping, in a motor home, trying to educate kids, trying to earn a living. So he's pissed at us being late, time has been wasted.

Gene lies charmingly about a faulty camera. Mundon cools down and wants to get on with it. I suggest we do some stuff in the trailer, you know, domestic stuff, see how these folk actually live between gunfights.

Gene with the Arriflex and me behind him recording sound, follows Mundon to the trailer. Mundon opens the door and steps in. He opens a small cupboard level with his face, takes off his hat, puts it in the cupboard, closes it, sighs. The hat is where it belongs, in its place in the structure.

He moves into the interior. We follow. It's anally tidy, nothing out of its place. Mundon raises the top of one of the banquettes where the children sleep to reveal his stash of weapons. No titanium barrels, they're all in good lethal order. He's telling us, with growing excitement, about each individual weapon. The .357 magnum pistol he uses to hunt elk. He demonstrates how he sits crouched, back against a tree, both hands gripping the magnum, getting a bead on the elk and then Pow! Rifles with telescopic sights are for sissies.

He opens up the other banquette, his ammunition stash, box upon box of cartridges for every one of his guns, .22, .25, .45, .38, .3030, .357 magnum. Mundon is a gun freak. His trailer is an arsenal.

Doris returns with the children, pretty eight and nine year-old girls, sun-tanned, long blond hair. Mundon's teaching them the fast draw. He's a man with his mind on the future, the entire family on stage.

In a field out back, Gene shoots a roll of film while the kids, pushing hair away from eyes, practice their aim with .22 Bearcat revolvers and live ammunition. There they are, two pretty little Californian girls standing in the sun, blazing away at targets and scoring. It's bizarre. I like it. Barry loves it. That's plenty for one day.

Gene's in a hurry to get back to the motel. I didn't know much about heroin but it turns Gene from morose to charming in a few seconds. All of us sometimes want to shift our mood from where it is to somewhere different, somewhere more carefree, I no longer care as long as he delivers.

The following day we're back with the Mundons out in the hills.

We're setting up the high-speed camera to film the thirteen-one-hundredths-of-a-second draw. We have the little white enamel camera loaded and set on a steel tripod. We're setting weighty sandbags around the legs. When this camera gets going we don't want it rocking about.

Mundon shuffles around in front of camera like an athlete getting ready, finding the right place for his feet. Gene sets the picture, focuses carefully on the hand that will draw. Everything's set, exposure, camera speed. Mundon places the polystyrene cup on the back of his gloved hand. Gene starts the camera, gives it a second to run up to speed and yells "Draw!!!" Mundon does the business. The camera winds down. It's difficult to gauge success or failure. We do it six times.

We hang around San Bernardino until Sunday because, in a meadow on the outskirts of town, the annual California fast-draw competition is taking place, a big day for the Mundons. Hundreds of motorised trailers are parked in ranks around the perimeter, tents are up, banners, flags flopping around, it's festive, everyone swaggering about in full cowboy uniform, Stetsons, boots, tooled leather gun belts, everyone toting a gun. This annual competition is important to Mundon. He's the champion and every year he has to prove it. It'll make a neat end piece to the movie.

But he doesn't win. He's pissed off, really angry. Gene follows him to his trailer, films him getting in and slamming the door on the camera. From inside the trailer, a muffled bellow of frustration, a different kind of end.

We drop off the film at the lab in Hollywood. They'll send the negative and a cutting print to us by express mail. We exchange cameras and tape recorder for editing equipment and head back to Santa Fé and *Confusion Cottage*.

There was no welcome at *Confusion Cottage*, not even a drink. Candice is alone in a dark mood letting off about Mike and his little cow wife. Barry gives ear to her bullshit about cocaine Mike sold her being cut. I'm dog-tired.

Alex's in our room watching television. She's pleased to see me. She spent a lot of time with Mike and Barbara, quiet time. They loaned her the TV. I ask how it was with Candice. It's been OK. Not bosom pals but OK. Candice doesn't like Mike and Barbara. Won't let them in the house. They were never invited in when they came to pick her up.

I tell Alex Gene is a junky. She's shocked, sighs, shrugs, says she's tired and gets into bed. I'm tired. I get into bed. We cuddle. I dream of Candice.

It's morning, late morning, warm sunlight streaming into the room. I'm in the kitchen drinking coffee. Gene's already up, cheerfully unloading the truck and carrying the editing equipment through the kitchen into the cutting room. He smiles, sings out good morning. I finish my coffee and help him. The Moviola editing machine, the old one-arm bandit, weighs a ton.

It's not easy between Gene and me. I've no moral objection to him using. I had to struggle with it at first but went with the flow anxious only about his dependability. Being possessed of a guilty Catholic conscience, he sees me, a non-user, as some paradigm of virtue and by implication a critic. I make him feel guilty just being there. What can I do about it? Nothing. I'm making a movie. I need Gene to cut and finish the film. If he needs junk to function, he can have it.

It's mid-day. Barry emerges. He and I sip beer in the kitchen, watching Alex preparing brunch, stacks of pancakes, crisp streaky bacon, eggs, strong coffee. She's usually particular about food, spends time on its preparation, pays attention to its nutritional value. She's not a macrobiotic ascetic, not at all. She's sensual when it comes to taste and texture. Barry likes either short order diner food or lobster and champagne, nothing in between pleases him. Today it's short order. Alex likes to please.

Barry is showing more interest in her, compliments her on her skirt, a long-to-the-ground soft cotton thing, black with flowers that she made from material she bought when we were living in Morocco. She looks good, tanned olive skin, sun-streaked hair down her back. She responds well to Barry's compliments.

The phone rings. Barry picks up, listens, hands it to me. It's Mike. He'll be down to collect us in an hour. It's quite a trick staying cool between the animosities of friends, taking neither side, letting them get on with it. It's OK with Barry, but with Candice our continuing friendship with Mike and Barbara is tantamount to high treason.

Mike's full of smiles when he arrives. He's thinner, more gaunt and rascally-looking, a red bandanna tied around his long blond hair. He seems confident enough, smiling as ever, but these days the smiles don't always reach his eyes. We embrace like bears. It's good to be with him again. We pile in the car. Barbara drives, Mike lights up a spliff and we're on the road.

We drive to their house the other side of town in a pueblo a few miles into the countryside where we sit relaxed sipping cold beer, listening to music, talking. It didn't come as much of a surprise. Mike isn't involved with film anymore, hasn't been for the past year or two. He's a full-time cocaine dealer. It provides a fair living, keeps them comfortable, he gesticulates around the room, everything they want. He's just bought the house and an acre and a half around it. He and Barbara take occasional trips to Columbia or Bolivia, buy a kilo, put it in his plastic leg and he walks it home. He smiles. I have to laugh at the karmic compensation for a road accident that happened all those years ago.

Mike unfolds a gram of brown crystalline powder. It looks like brown sugar. It's Mexican heroin. He lifts some to his nose on the long nail of his little finger, does the same for Barbara. He offers it to Alex. She passes. I pass. He smiles.

We sit silent as they drop down a gear or two. I'm thinking of the first time we met, Mike and me, in the flat in Fulham, how he'd greeted us, suited, drink in hand. Barbara says she wants to have a baby.

The film arrives back from the lab in LA. Barry is checking the invoice against the delivery note. All the cans are there. Great excitement as we open them up and lace them one after another into the Moviola. We stand around Gene seated at the controls.

The stuff's good, sharp of focus, excellent colour density. We watch the Mundons go through their stage act. The practice shoot-out gets a few laughs, then the stuff in the trailer. The fast-gun competition gets a laugh, Mundon retiring defeated to his trailer and his muffled anger as he slams the door gets a big laugh. We fall silent watching the slow motion of the draw.

I can't believe how slowly the drama unfolds, the inexorability of it. Mundon's hands move so slowly through the motion of fast-draw, the unsupported cup turns lazily in the air. From the tip of the gun barrel as it slowly rises to its target, a slow orgasm of an explosion, an ejaculation of flame and smoke, the cup disintegrating, fragments dispersing, drifting slowly to all corners of the screen. It's the most fascinating piece of film I've seen in a long time. Everyone's stunned. I clap Gene on the back. He grins. Even Candice knows she's looking at something unusual, something dramatic and beautiful.

Where's the stuff of the kids, the little blond angels blazing away? We haven't seen it yet. I check the lab invoice and delivery note again. All the film we left with them has been returned and accounted for. What's happened to it? Gene doesn't admit to it, says the lab has fucked up, but it doesn't take Sherlock Holmes to figure out he shot the sequences without any film in the camera and there was nothing to send to the lab. I'm pissed off but what's the point of accusing him? Where would it get us? Candice shrugs and goes to her room. I look across at Barry but can't catch his eye.

Living with Candice in the conjugal sense is a full time job. She's demanding of your attention at all times and because she's such a sexy bitch it's not too much of a problem. Barry has his hands full riding her mood-swings. The movie comes second. I'm not getting much support from him.

Gene and I are left alone to edit the movie. He's running through the practice shoot-out stuff in the hills, the Mundons facing off against each other. He's struck by an idea of how to deal with it. I ask him to tell me about it. He can't explain, says why don't I leave him with it for an hour, relax, have a coffee, he'll show me. I leave him for an hour. Editors sometimes adopt this kind of superior air thinking they see things others can't. They can't explain it to you, it has to be demonstrated. Relax. Have a coffee while I spend an hour of your life showing you what I mean.

Gene runs his construction for me. He's cut out all the 'business' and left a succession of takes of the Mundons facing off against each other, cut with increasing rapidity until it's one draw after another. What do I think? Not much. In fact I think it's corny commercial shit. I say,

"It's great, yeah, really good - but I'd like to try it another way".

He scowls. "What's wrong with it?"

I'm trying to keep it light.

"Nothing. It's really good in its own way. It's not the way I see it."

He's surly.

"So how do you see it Mr Director?"

I'm aware of the way this conversation is going but I'm determined to remain in control of the film, guard its integrity.

"You've lost the context. You've cut out all the interesting human stuff, the nose scratching, the uncertainties..."

Gene reacts to the criticism, his mood in swift decline. Without looking around, fists clenched, he snarls at me.

"Why don't you cut the fucking thing yourself wise guy".

He slams out, off to his room.

I want to cut the film myself but I'm not yet sure enough about procedures and the mechanics of it. I take a break. Neither Barry nor Candice are up yet. She likes to sleep on in a darkened bedroom with the warm smells. It isn't yet her time of day.

Gene doesn't show for the rest of the day, doesn't show for dinner either, emerging after we've eaten. He gets Barry off in a corner, voices lowered. Barry gives him some folding money and car keys. Gene leaves closing the door quietly.

Candice is expecting visitors, friends from Julian Beck's Living Theatre, a radical theatrical group. She'd lived with them for a few months in New York. They've been calling each other all day. Friends are driving them down here. They could arrive any time.

We're editing again. Barry's there and Gene's in a good mood. We find the best slow motion take and run it a few times. We all agree it should open the movie. It's one and a half minutes of film we're all happy about. We set up the shooting competition, Mundon losing and slamming the door. Gene cuts it together. We have a good laugh. We re-cut the face-off in the hills. Barry supports my view and Gene shrugs. He'll take it from Barry, Barry's keeping him high.

There are moments when I enjoy Gene's company. Away from the film and high, he's a pussycat, warm and affectionate, blaguing about life, his hopes and aspirations. He's charming when he's stoned. When he's not stoned, when he's run out of dope with nothing on his mind but scoring some more, he's insufferable, a pain in the arse and dangerous. Right now he and I are having an unsuccessful session in the cutting room. He's in a foul mood and won't take my direction. My patience is leaking away. The temperature's rising, confrontation is imminent. He stands, swings around on me fists clenched, eyes blazing.

"Get the fuck out of here or I'll knock you through this fucking wall!"

I'm scared. He's a big guy and angry as a bull. I do what he says. I'm losing control. I knock loud on Barry and Candice's bedroom door. Barry shuffles out in Candice's robe rubbing his eyes. I yell at him. He's the fucking producer! He's got to back up my position! He forces a smile, tells me not to worry. He'll be up and about in an hour or so, says he's sure it will sort itself out and wanders back into the darkened bedroom closing the door. Indifference hadn't entered our relationship before. There's distance growing between us. He's drowsy most of the time now, his speech slurred. He's sharing Gene's skag. Candice too. No doubt about it. Heroin rules.

I want to get out but have to finish the film. I want to get out but there's no money. I spent whatever I had getting here. When the film's finished, Alex and I will be on our way to California to take it to the lab for printing, that's the deal. The only way out with travelling money is to finish the film.

There never was much routine in the house, people feed themselves when they feel like it. Alex and I are eating dinner alone when the doorbell rings. I haven't seen the others for several hours. They're in a huddle somewhere, probably fixing in Candice's bedroom. So when the doorbell rings nobody responds. It rings again.

I go to the door. A bright eyed young couple bound in like we're having a party. They say "Hi". There are others with them getting their bags from the car. They're Candice's expected guests from the Living Theatre.

The young guy introduces himself and his girl friend. I'm Tim. This is Cheetah. He's looking curiously at me then at Alex wondering if we've met before. It's incredible. It's dawning on me, the dentist's office in Casablanca. Our memories collide. We laugh, hug each other like long-lost friends. Casablanca and now Santa Fé? What's happening? The others come into the house, the expected Living Theatre mob, three men and a woman. Tim and Cheetah introduce us telling them of the amazing coincidence.

They couldn't be any friendlier than they already are but are prepared to give it a try. It's a good omen. We all hug. They're elated, happy to be here. Where's Candice? Where's Barry? Tim's opening a bag of coke. It's party time, lights, music. Candice and Barry emerge in their robes, smiling. Greetings and hugs all round.

They haven't met Tim and Cheetah before. Candice is cool towards them. Asks them how they come to be here, says she doesn't remember inviting them. Cheetah explains in sexy French accent that it's their car. They are the friends who drove the Theatre people here.

If you think that's reason enough to be hospitable even to strangers, then you wouldn't be taking into account Candice's fierce defence of territory. The moment she set eyes on Cheetah she recognised her as a threat to her sexual hegemony. Imperious, she makes it clear sexy Cheetah is not welcome and wants her and her boyfriend gone from the house. Ultimatum delivered, she sweeps off back to her bedroom. That puts a damper on things. Barry shrugs and follows her. Tim and Cheetah are cool. We hang out with them for a while and then turn in. In the morning they and their car are gone. They leave no message. So much for omens.

The mischievous spirits of Confusion Cottage take over. Skag rules. For the actors it's just another party, just another scene in the grand play of life. Everyone except Alex and I are smacked out. Day is night, night is day. Drapes remain drawn. Shadowy figures around candles cook up and fix. Barry and Candice spend most of the time in their darkened bedroom with occasional visits from the others. Gene and the actress are engrossed in each other and seldom appear. Everyone's friendly enough but treat Alex and me like we're visitors.

Gene is more confident now he's part of the majority and getting laid. He's less abrasive when we meet, talks about the movie, says we must get down to work but nothing happens. Barry, out of the bedroom, drinks coffee with me, says not to worry. It's a party. He grins. We'll get back to the film as soon as everyone sobers up.

Then he lowers his voice. A trace of anxiety. He's running out of money, there's possibly not enough left to finish the movie. There's still a massive mortgage on the house and the bank won't lend him any more but not to worry. Candice has friends with good marijuana connections. They're thinking about dealing ten kilos, twenty if the money runs to it. Then off to New York. He's confident he can get top dollar for good weed up there.

I'm being asked not to worry quite often. We're getting into a marijuana deal to keep things going? What the fuck's going on? Barry's running out of money because he's spending it on skag. I'd never questioned the financing, took it for granted he had that side of things sorted. The movie sitting in the cutting room waiting to be finished is no longer the centre of the action; it's been relegated to sub-plot in a much larger movie. Real life has taken over.

I have plenty of time to think about things, plenty of plans but no resources. Mike and Barbara are no help. We haven't heard from them for weeks. Alex is phlegmatic. She wants us to get out but it's no sweat. She knows I've got to finish the film. I've got to get these junkies to move their arses.

I'm sitting on the toilet in Candice's bathroom. She's tying a silk scarf around my upper arm. She kisses me softly on the mouth like she'd do in New York days. Gene's cooking up some skag. Everyone's there, looking on, smiling. It's some kind of infernal baptism. Gene finds a vein. A sudden slackening of all resistance, a soft blanket descending on the mind, on the whole body until nothing's important, nothing's even worth thinking about. I'm stoned amongst the stoned.

Heroin strikes terror into the hearts of people who don't use it. It's dangerous, addictive, leads to degeneracy, death. It's largely true, a justifiable fear in modern life, children becoming addicted and all that. When Alex asks what it feels like. I tell her the truth. It's nothing, a bit of a rush, nothing.

Everyone in the Cottage covets and jealously guards their stash. I overhear arguments coming from Candice's bedroom about who's been stealing from hers, voices raised; accusations.

I'm back in the cutting room with Gene. He's cutting the negative, the one and only original, cutting it and splicing it together ready for printing. He's cheerful, friendly now I'm an initiate. He works fast with confidence.

Barry pops his head in, says the marijuana deal is about to go down. Candice is in her element talking tough. She refers obliquely to the deal as though it's a secret unknown to us. She wants the house cleared by six this evening, all of us out except Gene and the actress. Gene's needed. They're expecting some heavy players, bikers. They're coming over tonight. There'll probably be guns. For our own safety, we and the rest of the actors had better not be around.

No further encouragement needed, we packed a few things. Barry's booked us all rooms in town at the Half Moon where he and Candice used to stay before they bought the house. He drives us there. We don't see him again until the following day around noon.

He arrives looking wasted, says he hasn't slept since yesterday. The deal went down OK; twenty kilos; but after the heavies had gone they discovered the weed was wet, an old trick to increase its weight. Back at *Confusion Cottage*, the wet marijuana's in the empty swimming pool, drying out. Barry gets in there every hour or so turning it over with a gardening fork. Gene still hasn't finished cutting the negative. The house is low on dope, money's tight, Gene's in a funk. Everyone's coming down from the spree. The actors drift off back to New York where they live. The actress stays on with Gene.

Barry and Candice busy themselves preparing for the smuggling. As soon as the weed's reasonably dry, they buy a Sears Roebuck compactor to compress it into manageable bricks to be packed in their *Vuitton* luggage. They'll go first class by train. There's sufficient money left to take care of Gene's habit, long enough for him to finish the film.

Cowboy Jim arrives from Colorado. He's going to look after the place while Barry and Candice are gone, defend Confusion Cottage and its contents against whatever's out there with his .45 Colt automatic which he wears around the house in a shoulder holster.

Gene finally hands me the canned negative and sound tracks. He and the actress are leaving. Barry and Candice, bags packed, are ready. They dress elegantly and expensively for the part. They look wealthy and above suspicion. Barry hands me a hundred dollar bill and a brick of marijuana, a kilo tightly packed and saran-wrapped. He says it's worth a couple of thousand dollars. I should sell it to pay the lab costs.

Is this really happening? He and I take a long look at each other. I have to smile, the way things have turned out, the way the life has become the movie. He grins. Says we must do it again sometime. Candice is in a hurry and has nothing to say. Cowboy Jim drives them to the station. It's the end of something. I'm not sure what exactly, but it's definitely the end of it. *Confusion Cottage* changed me. It changed Barry, too.

Alex and I pack our bags. I call Bob in London to find out if Al ever got to LA.

He's there with his son and here's his phone number. I call. It's Al.

Sure we can crash with him.

Alex and I get out on the road with our bundles. It's easy to get a lift when you're with an attractive young woman.

Al lives high up in the Hollywood hills, one of those houses at the top, overhanging the rim of Laurel Canyon. Our 'lift' is kind enough to drive us up there. Al hears the car, comes out to welcome us and help with the bags.

We enter the house by the living room. There's a grass green carpet wall to wall, a white-painted picket fence around the kitchen counter. A sun-bleached trunk of a tree, stripped of leaves, twigs and minor branches, is set in the middle of the room, its base firm on the grass-green carpet, its naked limbs reaching up to the sky-blue painted ceiling. A number of cats sit comfortably in the tree watching us.

Al drones on about life in general, his in particular. He's been taking pictures of musicians, recording artists. The art director over at Columbia has become a good mate, gives him a lot of work, loves the pictures, a really nice guy, a truly spiritual person. I must meet him.

Where's Michael the kid?

God alone knows. Went out earlier, no idea when he'll be back, hope he's not up to mischief. There'd already been a tangle with the LAPD.

Got anything to smoke?

I take the kilo out of my bag. Al faints and recovers like people do in movies to express their incredulity and unbelievable joy. I break it open. He rolls, we smoke. He shows us around the house, where we're going to sleep, bathroom and all that. Everything's simple, clean and tidy. Alex sets about unpacking, getting the room the way she likes it. I leave her to it.

The veranda at the back of the house juts out over the canyon facing west into the setting sun. Al and I sit in its warm roseate light watching it all the way to the horizon. We sip cold beer, happy to be in each other's company again. It's moments like this that past lives flit through the mind; the studio on the Lower East Side, Al the fast lane art director.

Alex is cooking when Michael turns up. The door slams and there he is, fifteen and tall as Al. He moves awkwardly without rhythm, doesn't smile, eyes lowered. He's tense, short fused, violence close to the surface. When we're introduced, he grunts. Al's obviously used to it and makes pleasant conversation with him as though he's a normal sociable person.

Where's he been? Did he have a good day?

Michael says, "Aaaaagh".

We eat, wash the dishes, smoke and listen to some music. Michael's asleep on the couch. Al swings himself into a hammock under the tree, a miniature TV on his stomach and smokes a joint watching a movie. It's been a long day. Alex and I turn in.

We don't have a car. The only way up and down the canyon is to hitch-hike. Sometimes Al borrows the spiritual art director's car to do the shopping. He borrows it to take me to the lab with the film. They'll let me know when it's ready.

We don't have any money. So what? There's nowhere we want to go, nothing to do but relax with Al and the cats. *Confusion Cottage* took its toll of us.

A phone call from the lab. They're apologetic. The negative fell apart on the printing machine. Worst splicing job they've ever seen. They can take it apart, clean it up, reassemble it, re-splice it, but it'll cost. Will I get in touch with the producers, get their instructions? They must be joking. After what I've just gone through? Producers? Instructions? Fuck them! Fuck the film! I'm overwhelmed by anger and frustration. I want to be shot of this movie forever. I give them Barry's number in Santa Fé. Let him deal with it if they can find him.

I'm thinking about the Swami. When my mind and emotions get in a twist, it calms me to think about him. I always carry his photograph, one I took. It's propped in its leather frame on the windowsill in our bedroom. Every morning after a shower, I go patiently through the long sequence of asanas I learned from him. I'm slowing down to walking speed, paying more attention to the grasses and flowers growing at my feet, more attention to the beautiful Alex. She's good company when I slow down enough to appreciate her. We eat healthily; rice, fresh vegetables, fresh fruit. Even with such virtuous raw materials Alex manages to titillate the tongue, get the saliva flowing. All of us appreciate it, even Michael.

The next-door neighbour's giving Al a hard time and not mincing words. He owns two fluffy white thoroughbred Persians, princesses. One of Al's cats, a large black, ill-tempered tom named Croak, has been harassing them, spraying everywhere. The neighbour doesn't want some scruffy alley cat jumping his princesses, he has classier suitors in mind; thoroughbreds go for five hundred a kitten. He's lodging a formal complaint.

Al discussed it with the rest of the cats. They were unanimous. For the sake of social harmony, Croak's balls have to go. I understand it was the same clique voted to throw out Goober, an undistinguished, shorthaired, uniformly grey chap with yellow eyes. Al says he somehow reminded him of Michael. Symbolically he wanted to be rid of him. He put Goober on a plane to Philadelphia, sent him to some people he knows. They promptly put him on the next plane back.

Al's big on social harmony. He's probably never voted for anything in his life but you could say he's a communist. He reads a lot about Mao, identifies with the struggle. In the Korean War, he was seventeen years old lying in his shallow foxhole, barely in hailing distance from his nearest mate. He watched the treeless horizon, from East to West as far as the eye can see, filling with thousands of Chinese soldiers. They advanced line abreast, bayonets fixed, bugles blowing, thousands upon thousands of them, an engulfing tide. Closer and closer, he could see their faces. They swarmed over him, shot him, left him for dead.

He still bleeds for them, feels guilty for the imperial injustices they've suffered. Can you believe it? He wants to travel to Viet Nam,

Korea, China; he wants to photograph the heroic struggle, become part of it. Sometimes he's difficult to figure.

This afternoon he drove up in a gleaming sunshine-yellow Volkswagen Beetle convertible. He's had a lot of jobs these past couple of months, accumulated more dollars than he can count so he buys this car thinking it will open up new horizons for him and the kid.

Michael wants to get it on the road right away. They can get out, go places; he doesn't care where as long as they're speeding down some highway. He admires the car, walks around it, kicks its tyres. He could drive it.

Al's never seen the kid keen about anything before. Here's a chance to improve relations, that's all he really wants to do, get closer to the kid. He wants to be an indulgent father of the kind he never had.

He asks him,

"Where you wanna go Mike?"

"Aaaaagh... anywhere!".

"Let's go up the coast to San Francisco. We'll visit with Zeke and Venus; I'll show you around the place, we'll have a time of it."

Michael grunts in a pleased kind of way. True existentialists, they don't pack a bag, not even a toothbrush. They pile into the car that very moment and drive off on their five hundred mile journey.

Alone, Alex and I are settling back into our routines, Yoga, meditation, vegetables and fruit. I've become healthier than I've ever been, body and mind never more relaxed. I'm making a few notes, an idea for a movie. Alex is feeding the cats, feeding us all.

We're eating dinner at sunset on the veranda. The telephone rings. It's Al. He's calling from the hospital in San Luis Obispo not a hundred miles up the coast. They've had an accident, spun out on the soft shoulder and overturned. The car's totalled. They're all right, nothing serious, a few grazes. They'll hitch a ride back and see us sometime tomorrow.

Al shares the fortunes of Job. He's well meaning but things rarely go right for him. He's easy, but nothing else is. He doesn't complain but you see it in the stoop of his shoulders, his shuffling walk. He's apologising for all the sins of the world. You want to put your arm around him.

It's late afternoon and hot. Local dogs are kicking up a din. I go out the front door to see what all their fuss is about. Al and Michael are limping wearily up the dusty road, a mess of torn clothing and bandages, a small, hairy dog yapping at their heels, snapping at a loose end of bandage from around Michael's ankle. He kicks at the tiny fellow. All the neighbourhood dogs in their gardens are hooting and hollering, welcoming them home.

We're waiting for Barry to arrive. He phoned last night. He's driving across from Colorado. He'll be here this afternoon. There's a change in the atmosphere, an edge of expectation, Barry's coming.

You'd expect me to be circumspect, cautious, maybe, after what I went through with him. But I'm looking forward to seeing the bugger again. He's coming alone. What's happened to Candice? Who cares? There'll be a lot to talk about when he gets here; it's been more than a year. Alex is prettying herself, lipstick, a little make-up, the first time in I don't know how long.

Barry drives up in a brand new Wagon, same Barry, same shotgun hanging in the back. He brings it into the house with his bag. The gun impresses the hell out of Michael. Barry checks it's unloaded, tosses it to the kid to play with. He shoots at the cats, muttering how lucky they are he's out of ammunition.

What about Candice?

Candice? She's an all-time bitch and can go fuck herself. They got the grass safely to New York and sold it for top dollar, twenty thousand. She takes fifteen and gives him five. Says she set it up, paid for it out of her own account, it's her deal, money talks, bullshit walks.

He went back to Santa Fé, sold the house back to the mortgage lenders, barely breaking even. He returns to England and brings Jacqui and the boys to Pagosa Springs where he left them yesterday with Jim. He's stony broke, still owes on the wagon parked outside.

Everyone crashes except Barry and me. We sit in the dark on the veranda, drinking and smoking. He asks what happened to the film? Do I have a print? I guess he's hoping to make some money out of it. I tell him what happened. He shrugs, says it's some kind of ending. We laugh and drink to it, but he was hoping.

He's read in a paper, a magazine, the movie I wrote has been released all over. Have I heard anything? Aren't I due a share of profit? I hadn't given it a thought. In the morning I call Twentieth Century. They say they sent me a cheque two months ago to the address on the contract, the flat in Fulham, London England.

"How much?"

"Fifteen thousand dollars."

"Fifteen thousand dollars?"

I tell them to stop the cheque. I'll be right down to sort it. Disbelief and excitement. I'm stunned. Sitting here for more than a year, a virtual recluse at peace with myself and the world around me while the mills of God were grinding away slow and sure, sorting out the next move, the next chapter. Alex puts her arms around me, smiling, eyes shining, happy as a child.

The money changes everything around. It terminates our monastic existence, our retreat from the world and puts us right back on the road, smack in the middle again all in the twinkling of an eye. We have options.

I get the money from the studio. I give Al two thousand dollars back-rent. Smiling, he's embarrassed to accept it. He's going travelling, taking Michael with him to Vietnam. I give Barry a thousand, Alex a thousand. There's excitement in the air. Everyone's packing, saying goodbye, on their way to the next phase of their lives.

The three of us, Barry, Alex and me, are driving to Miami. From there, we're going to fly to Paradise Island in the Bahamas to a house on the beach belonging to wealthy banker friends of Candice. It's unoccupied and Barry has the keys.

We're driving east on route ten, straight across the Southland, undisputed redneck territory. Barry and I are longhaired so have to be circumspect. We don't shave for a couple of days, wear dark glasses and look mean and funky. Alex sits between us. Barry scatters shotgun shells on the dashboard to give added meaning to the gun lodged in the back window. We scowl when we're filling up in gas stations and make it safely to Florida.

We stop the night in the Tallahassee Hilton. I've turned in. Barry and Alex are taking a midnight swim in the pool on the roof. In the morning we'll drive south to Miami. Tomorrow night we'll be in Paradise.

We take the flying boat from Miami and a taxi from the mooring where it sets down on Paradise Island. After a ten-minute drive over uneven roads tunnelling through dense sub-tropical vegetation, we arrive at the beach and a large, elegant Victorian beach house. Beyond it, white seams of surf break quietly on the soft sand. It's completely isolated, not another house in sight. The air is warm, the surf the only sound. It really is paradise. We move right in. Barry says there's a restaurant a mile or so down the beach so we don't unpack, we dump everything.

Daylight is fading as we set off; it's dark when we return. Tired, we stumble around looking for light switches. We pick our bedrooms. Alex is behaving strangely, brow furrowed as if there was something she'd forgotten to do. I ask if she's OK. She doesn't answer for a moment, then turns to me smiling shyly, says she's going to sleep with Barry. Jealousy is an emotion I'm familiar with. It's a canker that eats away at your very soul. My marriage along with a few drug-inspired insights into human nature had inured me to it, cured me of it so I thought.

I'd never heard these words in Alex's mouth before. They don't belong there. I feel estranged by them, uncomfortable, shut out. I shrug, grin and say it's OK. She knows it's not. She kisses me and leaves, closes the door quietly. I have no proprietary claim on Alex. Her attraction to Barry had been obvious for a long time. I'd not given it much thought. We were all such close friends it didn't seem out of place.

The following day the weather turned. Barry and I walk along the overcast beach, the two of us, my mind desperate to make a plan that will release me from this feeling of exclusion. I puff on a joint.

Barry's not phased at all by the change around. He behaves perfectly naturally as if nothing has altered between us. Has anything altered between us? I pass him the joint and relax. We sit on the sand listening to the surf and sea birds. Any tension, real or imagined, dissipates. Barry embarks on a story that a poet he used to publish told him. It was when he, the poet, was up at university reading English Literature. He and a mate develop a strong antipathy towards one of their lecturers, a pompous arse. They invent a character, a woman, to enter into correspondence with him and, through this alias, make known their feelings about him and his work.

As I listen a movie begins to take shape in the mind. Barry's words recede as the imagination grabs the idea. I see the whole thing, the plot, sub-plot, beginning, middle and end, the strangest of love stories. I leave Alex and Barry on the Island. They bring me to the flying boat. We hug and kiss like the friends we are. I'm returning to London. I have work to do.

104

I'm staying with Bob again. Kelly has been moved to hospital in a final attempt to save her. I've been here for two months and done little work. The new story is ever on the mind but I don't get around to writing it down; the mood in the flat isn't conducive. We visit Kelly in the hospital every day. She hardly says a word. It's too much effort for her to speak unless she has to. They'll send her home when nothing more can be done. The cancer has its claws deep into her.

Candice is in London. She comes to see us. She and I are alone in my room. She's friendly. Today I'm her confidant. I listen sympathetically to her tales of woe; her break-up with Barry accusing him of infidelity with some bitch in Santa Fé protesting she's not that kind of woman, always loyal to her man. She'll never forgive him.

I don't interrupt, I don't tell her what Barry's been up to with Alex, that would spoil the empathy building between us. Listening to her voice, being in close proximity with her again is giving me a hard-on. She says there's a young coke dealer in New York trying to get into her knickers. He's wealthy but just a kid. She flew here to get away from him.

She says I look stressed. She's concerned. I ought to take better care of myself. I'm hers and she knows it. She pulls a small box from her bag, opens it, says she's got something to make me feel better. I've never felt better in my life. An expensive, stainless steel and glass syringe, already loaded with a clear fluid. She ties off my upper arm with her scarf, flicks at a vein, shoots me up. The soft blanket descends. She's kneeling between my knees unzipping my fly.

Late morning. We were in bed asleep. I wake at the sudden opening of the door. A distraught young guy stumbles in, says he's looking for Candice. He sees her next to me, freaks and runs off. I can hear him crying to Bob and Bob trying to cool him out.

I get dressed and join them in the kitchen. Bob's frying up. Like Barry, he has a fondness for the short order. He's cooking up stacks of pancakes and crispy bacon. The young dude sits miserable with a cup of coffee. Bob introduces us. His name is James. He's just flown in from New York.

James can't bring himself to look at me. Uncomfortable just sitting at the same table, he goes to the toilet. Bob tells me James is an occasional business partner he met through Candice. The kid's so confused and distraught he's just handed him the keys to a Land Rover, ownership papers, everything. He'd taken it as trade from some cokehead filmmaker but can't be bothered to pick it up. He'd got no use for it. It's sitting in Ibiza. I hear the toilet cistern flush.

This is Candice's ardent suitor? Bob nods, fiddles with the Land Rover keys. We hear muffled murmuring coming from my room, the door opening, James saying, "See you later". The front door to the street opens and closes. He's gone. Bob tosses the keys on the table, looks at me, shrugs.

Candice cool and composed tells us she's going to marry James.

She has a weakness for slaves, the richer the better. He's a rich slave; he's just given her five thousand dollars cash to buy something to wear for the wedding. She's meeting him in Paris. She carelessly discards my adoration in favour of this squirt of a coke dealer. I'm seeing myself bobbing around in the flotsam of other people's lives, a sometime bit part player in Candice's saga, an ineffectual onlooker in Kelly's struggle to stay alive. I'm not writing, not doing anyone any good, least of all myself. I'm at low ebb when Alex calls.

She's been back a month. She and Barry parted a month ago fuck him. He went off to New York looking for Candice. Alex has changed. She's talking. Barry unlocked something. She's staying with one of her sets of godparents down in Wales, the Forest of Dean, the Wye Valley. She says it's beautiful this time of year. Why don't I come down? I've just had another cheque from Twentieth Century bigger than the first. I buy a car and set off for Wales.

The godparents are hospitable gentry. Prudence is an ex lover of Alex's father, Jasper is a warm and friendly Quaker. They live in a large house overlooking the Wye in the middle of a few acres of cultivated land. Prudence is never satisfied. She thinks their ten-room house too small and bemoans the time when they owned Jasper's family mansion and extensive estate. But darling Jasper played the horses and now they're down to this. She's a silly woman but Jasper adores her. No children of their own, they treat Alex as a daughter and she's happy with that.

The countryside welcomes me with its clean air, bird song and distant mechanical agricultural sounds pleasing to the ear. I could live in this kind of environment, I could write here. I ask Prudence if she knows of any properties in the area to let. She says Mrs Davis. She's on the phone immediately. Mrs Davis says that one of her properties became vacant this morning. We arrange to meet her there. Jasper knows where it is; he'll drive us.

We leave the main 'B' road and bump along farm tracks, turn onto a narrow lane that descends into a valley. Traversing the side of the valley we pass an occasional house perched on the slope then nothing but a vista of pine-forested hills. There's a splash and gurgle of water. The lane flattens out alongside a stream spanned by a small stone bridge, a stream that had once driven a mill's water wheel. Twenty feet back from the bank, in front of the ruin of the mill, the miller's cottage.

I've lived in many places but never before had the feeling when I first set eyes on them that I was coming home. Mrs Davis's quickly shows us around the house. Telephone, electricity, heating, a kitchen large enough to eat in, Aga stove, slate floor, a carpeted, low ceiling, living room, ingle-nook fireplace at one end, bedrooms upstairs in the eves of the house. It's totally isolated and surrounded by hills and forests. Mrs Davis asks me how long I'm thinking of staying. I say, "The rest of my life" and give her a cheque for the next six months of it.

Of course Alex moves in with me. She has her bedroom and I have mine. Mine, at the front, overlooks the stream and the bridge. The moon is full the night we move in. I watch from my window a male fox with magnificent brush drinking at the stream. Alex's room is at the back of the house a good distance away. Apart from sleeping separately we carry on as we've always carried on. I write, she cooks and takes care of the house. But something is notably different. She talks more, even initiates conversations; spends more time on clothes and make-up; affects a different walk, more languorous like Candice. She's not hiding anymore. A newfound confidence is flowering. Saturdays we drive a dozen miles into Monmouth to do the shopping. It's become a regular affair ending up in the George, a pub where the gentry, mostly up from London for the weekend to air their second homes, meet each Saturday lunch-time to enjoy saucy gossip, get drunk and flirt with each other's wives. They're amusing enough. Alex knows them all. An idea for another movie is beginning to form in the mind demanding my attention.

After leaving Barry and Alex, on my way home to London from Paradise Island I spent a few days wandering New York. A book, stacked and featured in a store window, attracted my attention. *The Politics of Heroin In South East Asia.* I bought a copy. It's a well-documented history of the growth of the heroin trade from its legal sale in 1901 as an over-the-counter analgesic pill to a mammoth illicit industry with economic and political power enough to affect the course of history. The authors describe the Viet Nam war and much of South East Asian politics as a condition of the opium economy. In the middle of the war, army generals commandeered planes to fly raw opium to heroin factories in France and China; the CIA, operating unfunded, without their government's sanction, were buying and selling opium to raise money to buy guns for the local peasants to defend their way of life against the advancing Communists. True or false, it's plausible. I draw on other popular notions of power and corruption, notions of criminal conspiracy, the attack on Pearl Harbor perhaps, certainly the assassination of President Kennedy. I link them, insinuating them into history as the machinations, plots, stratagem of a singular man of power. Not the Mafia, someone more powerful, someone of exceptional wisdom and credibility who has the ear of Presidents and Kings. I see him. William Harper II enters my mind complete with genealogy. I sketch it out. It looks good. I'll deal with it when I've finished the love story.

Alex has a lover, a handsome young gentleman farmer, darling of the local gentry. He has a wife and family and a reputation for philandering. Everyone jokes about it except his wife. He comes down to the Mill around midnight or later, after his family are all safely tucked up. I hear his car's loose exhaust clattering down the rocky lane. He's shy around me, doesn't understand or, perhaps, believe that my relationship with Alex could be Platonic. When he visits, he avoids me and goes straight to her room. The affair doesn't last long. He's not at his best trying to make love to her with me in the house. Anyway I'm far too busy to pay him any attention. I'm writing a synopsis of the love story.

I've been introduced to Linda Seifert a young lawyer who prefers the glamour of show business to the dour, predictable drab of the courts. She's an agent for screenwriters, interested in me because of the success of *Vanishing Point*. I send her the synopsis. She calls, says she's read it, loves it and agrees to represent me. She'll show it around but warns me we'll not get much further until it's a screenplay.

Kelly's been released from hospital. I'm on the phone asking her how it's going. Her voice is thin but confident. She says she's dying. Just like that. I'm choked, can't talk. She asks when I'm coming up to town. I say soon.

Her funeral is a great social event. It makes the evening papers. Bob is calm. As much as he loved her, her death is a release for both of them. He does what has to be done, arranges undertakers, crematorium, covers all the other tasks and duties that fall on the shoulders of a bereaved widower.

The crowds of people that attended were spontaneously drawn, mostly magazine people, editors, models, hairdressers, fashion designers, photographers, a few musicians from famous rock and roll bands. There weren't any invitations. Word gets around. All of them know her, have a personal affection for her, a colourful crowd paying last respects to one of their own. Marvin Gay is singing Let's Get It On.

Bob packed a bag. He's coming back to Wales with Alex and me. It's night before we get on the road, late when we arrive at the cottage. Alex goes to bed. Bob brings out some morphine ampoules, the remains of Kelly's first-aid kit. Kelly is released. Bob is released. I join them, sitting back, gazing into the log fire, a confusion of thoughts and feelings drifting like vapour up the chimney dispersing with the wood smoke into the night.

Bob's gone back to the city to attend to business and generally get on with his life. I'm restless. The winter's been good. I enjoy the cold, moist climate of these hills and forests. I work well here but it's time to break out. I want to get on the road again. It's time to feel the sun. Alex's happy to take care of herself so I'll go back to America for the summer.

I spend a weekend up in London with Bob. He'd just heard from Barry and Candice, they're back together again. This time they're going to get married.

"Married? I thought she was already married. What happened to James the coke dealer?"

"Yeah. James." Bob shrugs, "She says he OD'd on their wedding night."

Bob produces a bottle of Jack Daniels and glasses. He pours; we toast the departed soul of the young coke dealer. Bob continues. She's a rich widow of course but she has to go and get it. It's in deposit boxes in a dozen different cities. She made sure she had the boy's keys before the embassy collected his body then met up with Barry in New York and together they went gathering. They're back in Santa Fé. He has their number. .

"Did she kill him?"

"Ask her."

Without knowing what I'm going to say I call her.

She's friendly. I commiserate about the untimely death of her husband and congratulate her on her engagement to Barry. She thanks me oblivious of the sarcasm; they want me at the wedding as Barry's best man; wouldn't be the same without me, we've been through so much together and so on. I can tell she's on a downer of some kind. She gets sentimental on downers. She gives me the date and directions how to find them, a ranch outside Santa Fé.

Shortly before I leave, Alex has a visitor, a striking-looking Dutch girl named Taetske who she'd met at a party up in London. She's tall, slim, with a head of wild hair, high cheekbones and flared nostrils. She'd been in love and living with a young American who'd been living living and working in London. He'd fallen ill with cancer and she'd nursed him until he died a few weeks ago. She was feeling empty and not knowing what to do with herself. She didn't want to go home to Holland. She has friends in California; maybe she'll go there. Heading that way myself I offered her a ride. A few days later we fly together to New York.

In New York I buy a second-hand, powerful and sleek v8 convertible. We get a pup tent, sleeping bags, a butane stove, stow them in the trunk, fill up the tank and drive north into Canada.

There's nothing quite like the sense of liberation a full tank of gas gives me. My heart is light as a feather and I have a new friend. I drive wherever we choose, wherever our fancy takes us, stopping where we like, a lake, a view, smoke a joint, sip a cold beer. At night we cuddle up in the tent and make love. We were not in love but at that moment in our lives we both needed it.

We drive around Lake Ontario and west across the Great Plains.

Turning south, we re-enter the United States, heading for Montana where we had friends. The US border guard stops us, ferrets around inside the car, the ashtray, the floor, between the seats.

"What's this?"

A young blond grim-faced uniformed border guard is holding up a marijuana seed between thumb and forefinger for me to see. I take off my sunglasses, peer at it mystified. I'd warned Taetske not to roll joints in the car. Anyway, there we were, confronted by an enthusiastic lawman accusing us of being responsible for this marijuana seed.

A friend, a Scientologist, once told me how to reverse the energy of a conversation. Always answer a question with a question. I give it a try.

"I dunno. What is it?"

He's impatient and angry.

"It's a marijuana seed!"

I plead innocent.

"Never smoked anything in my life. I don't know how a seed got in the car, maybe the previous owner?"

He's not having any of it. He knows I'm a dope smoker. I fit the FBI's profile of the typical doper and he's going to prove it. Taetske is hauled off into another room; he orders me to strip.

I'm handing him my second boot for examination. His hand gropes around in the foot of the boot, his eyes light up. He holds up this tiny plastic bag containing about three grams of weed. He's triumphant. I give up. Three grams of marijuana makes fools of all of us. If he hadn't found it, I'd be down the road whistling and he'd be the fool. He finds it and I'm the fool. I ask what happens now. He says if this had been Montana he'd be handing this fool over to the sheriff.

Montana? "Where the hell am I?"

"North Dakota."

He impounds the car on account of the importation of an illegal substance and I have to give him a hundred dollars to get it back. I pick up the keys, a shake of the hands, no hard feelings and Taetske and I are off.

We continued on down to Pagosa Springs in southern Colorado where Barry had settled Jacqui and the boys. It's still the small cowboy town I remember, decked out for a July 4th rodeo. We drive along a track to the outskirts, past a timber mill, across a broad meadow to a cottage, small and isolated in a vast Rocky Mountain landscape. And there's Jacqui coming out of the house; I guess she heard us. She can't believe it! I park, jump out and hug her tight. We haven't seen each other since the advent of Candice in her life. I introduce Taetske. I ask about the boys. The young one's gone fishing with Jim; the older boy's inside doing his homework.

We all eat dinner together. Whatever's gone down between their parents, the boys seem well enough adjusted. After they've gone to bed and Taetske's turned in, Jacqui and I sit out on the veranda enjoying the cool night air. She's been patient, hasn't said a word about Barry and Candice, now that we're alone she lets it all out. She's bitter. Not just about them, about things in general. How could her life have changed around so abruptly without warning? We were all having such a good time, weren't we? She avoids mentioning Candice. Barry's told her about the wedding. She asks if I'm going. I nod. She stares off into space. I put my arm around her. She nestles into my shoulder, asking the night what she did wrong.

Today we went to the rodeo. A cavalcade of horses and riders, men and women, boys and girls, flags, bands, local politicians and dogs, all jogging through the little western town led by the Rodeo Queen, a cowboy vision of busty female beauty. We followed them to the rodeo ground.

Wild-eyed horses held to rein, strut majestically into the arena, riders saluting the crowd of cheering, waving townsfolk. They line up in front of the town dignitaries; the band strikes up Oh Say Can You See. Riders sit, heads bowed, Stetsons over hearts, flags flap in the light breeze. The mind's camera is turning over.

Remembering the way Candice had treated the French girl Cheetah, I knew the beautiful Taetske wouldn't be a welcomed guest at Candice's wedding so she's heading for her friends in Santa Rosa in Northern California. We say our farewells.

I drive the now familiar route south from Pagosa Springs to Santa Fé. It passes close by Mesa Verde where ancient culture has been in conflict with industry for decades. The Peabody Coal Mining Company mines open cast on land sacred to the local Natives. There have been incidents that have come to blows, shots fired. The West is still wild.

I use driving time to ruminate on the new project. When in the middle of a story that's mostly what I do. Ruminate. If asked what am I doing with my life I'd have to concede I spend large amounts of it in my head, not exactly oblivious of what's going on around me but certainly preoccupied with realities of my own invention. I was never very good at small talk, social gatherings. When people wander up to me cocktail in hand curious to know who I am and what I'm up to I quickly turn any conversation to the story I'm writing at the time. I don't apologise for it. As I'm driving down to Santa Fé, in my head the Japanese are bombing Pearl Harbor, the President is declaring war on them and their Nazi allies.

I follow Candice's directions to where they're staying; a run-down working ranch with a large bungalow to the side built and rented out to help pay the ranch's rising costs and diminishing returns.

They're both relaxed and happy; I've never seen them like this before. Maybe love has won out. I notice Barry wears an ear decoration, a delicate pendant of silver and jet dangling from his right lobe. Candice, wearing its partner in her left lobe, says they're from the grave of a Zuni chief. They show me around the place, introduce me to their hosts.

The ranch is owned and run by Lucy Manning, a robust woman in her sixties, and her two sons Biff and Brad. She's eccentric, collects snakes, keeps them in a wire-fronted hutch feeding them rats and mice she catches around the place. When she was a younger woman in World War Two, she ferried newly built military aircraft from factory to operational airfields. Now she rides a motorbike with sidecar.

Her older son Brad, about twenty-eight, is married to a pretty young Swedish blonde and currently out on bail on a murder rap. He killed a man who had the nerve to talk to her. He's walking down the main Street of Chimayo a little town up in the mountains, sees his pretty little girl wife sitting chatting with a man in a car parked in the middle of town. He walks over to the car. No questions asked, not a word, he just blows the guy's brains out all over his wife's lap, just like that, through the car window. It's a *crime passionel*, everyone agrees. No one, not even the judge who bailed him, thinks badly of him for it.

His younger brother Biff on the other hand is a child of nature, wouldn't hurt a fly. He cares for the horses and grows a patch of marijuana.

Early morning. Shadows still long and the air cool. Barry and I ride out across the ranch. Barry's a fine horseman. He learned when he was in the army, sensed it would one day be a useful social accomplishment. I can't ride at all, not really ride, never learned. I still can't balance on a tight-arse English saddle but sitting in the wide expanse of a Spanish work-saddle on a cow pony I'm reasonably secure. We canter for a while. Then, ambling along, two cowboys of old, we smoke a spliff and chew the fat.

I told him about Pagosa, how I'd seen Jacqui and the kids, the whole crowd. I tell him about the rodeo. Rodeo? He muses for a moment then says there's a rodeo next weekend over at Window Rock, the Navajo reservation. We should go. I agree. After a respectable pause he suggests we film it. We've got time to kill before the wedding, a month or so. He looks at me, then at his fingernails, waiting for a reaction.

A film of Indians doing the Cowboy thing? Sounds interesting; another curious dish of Americana. If things got out of hand as they did in *Confusion Cottage* I could always walk away. This time I have money. We shake on it. We'll make the film ourselves. We'll be the cameramen. I've learned enough from assisting Mike and junky Gene to be able to cut and finish a film. Barry tells Candice the plan. She's amused and opts out. We nod our satisfaction.

105.

Neither of us is familiar with the machinery of a movie camera so we recruit two assistants in Santa Fé, film students who know how to load and take a light reading. The four of us descend, fully equipped, on Window Rock, a fake BBC sticker on the windscreen. Barry has an Arriflex, I have a Mitchell slow motion camera, not slow like The Fastest Gun but slow enough to let you see the drama of wild-eyed horses and bulls bucking and arching to unseat their riders, every flickering muscle, flecks of foam, pounding hooves, explosions of slow-rising red dust, riders curling slowly out of saddles.

Early morning on the big day, contestants emerge from their motor homes stretching, yawning. New arrivals, RVs towing horseboxes plough the dust. Men practice with lariats, looping them again and again over the pommel of a saddle set on the ground like it was the horns of a steer.

We filmed everything in synch with sound picking up the detail. Fingers working oil into squeaking leather rope that will bind a rider to a bull, strapping on spurs, curses muttered under the breath, a guzzle of vodka for the courage. The rodeo follows exactly the same ritual I'd seen in Pagosa, parades, bands, Rodeo Queen but all are Navajo, even the khaki uniformed cops with mirrored sunglasses.

The place is packed out, hardly a white man present. Not physically present but he's there all right. These folk aren't Indians anymore, they're Americans celebrating an American holiday in an American way. It must confuse the shit out of the old folk, tales of Custer's defeat at Little Big Horn still alive in their minds, watching their kids preparing to ride bulls and broncos, swilling vodka, strapping on chaps and spurs, standing silent, hand on heart while the band plays *Oh Say Can You See*. I chat with a Brave. He's friendly enough. Invites me down to Che Canyon, his dad has a gas station there. He punches me playfully on the shoulder, says invitingly, "Bring a crate of beer, man, and we'll kick the shit out of each other."

The fake BBC sticker gains us entrance to the heart of the matter, the rodeo arena. We filmed there all day, Barry catching the events in real time on the Arriflex while I covered them in slow motion. It's dangerous in there, wild horses, Brahma bulls all over the place. It's difficult to register their proximity, looking at them through a camera viewfinder.

Night. Away from the rodeo ground, Braves near naked, bodies and faces smeared with paint, dance ceremony around a blazing fire, the sounds of their drumming and singing rise with the sparks from the fire into the night sky to join with the spirits of their ancestors.

Making the film is easy. Since Barry's forever at Candice's beck and call, I have only myself to please. We set up a cutting room in one of the unused bedrooms and, as soon as the film's back from the lab, I get to work on it.

Seeing the rushes, the movie in them is obvious, the beginning, middle and the end. I match rides shot in real time with the same rides shot in slow motion. I find a place to cut from the frenetic pace of real time, a rider bucked this way and that, hanging on for dear life, to the unworldly, dream-like quality of the slow motion.

Rides are interspersed with preparation, the chaps, the spurs, the ropes. Riders mount the angry animals already struggling and kicking against the chutes, they tie themselves on, signal the steward to open the gate, then fly out like cannon balls trying to hang on for the eight-second ride. Few make it. Eight seconds on a crazed male animal, a leather cinch tied tightly around his balls, is an eternity.

Barry knows a group of musicians in town that play acid blue grass music. A guitar, fiddle, bowed cello and banjo. The fiddler is from New York, a burned-out child prodigy. He plays blue grass with the technique and imagination of a Master.

106.

We show them the rough cut of the movie and they start playing, upbeat and easy-going to the early morning scenes of preparation, fast and furious as riders and animals catapult out of the chutes and buck around the arena. When the slow motion cuts in, the music becomes strange, ethereal, violin in squeaky cadenza, sporadic chords plonk from banjo and guitar, undertones from the cello follow wild-eyed horses stomping the red dirt of the arena, bucking riders slowly out of the saddle, out of control.

One of the riders isn't hanging on in desperate fashion. He goes with the rock and roll of it, allows his body to find the rhythm of the wild thing he's riding, grinning all the while like he's on some kind of high as he's flopped around in the saddle.

It ends at night with the ceremonial dancers stomping around the fire, hollering and hooting to the drumming, fade to black and that's it. I synch up the sound, cut and splice the negative and draw up a dubbing chart. I'm taking the film down to LA for dubbing and printing. Candice promises not to start the wedding without me.

Jilson is a girl. When I get to LA with the film I've nowhere particular to stay. Al and Michael are still in South East Asia, somewhere in Cambodia, China, who knows where. So I call Jilson, she invites me over, has a spare room. Her name's Jill but she prefers Jilson. We met when Alex and I were living up in Laurel Canyon with Al.

Jilson is a groupie. Rock and roll is her religion, intoxication her preference over cold sobriety. She's swaying a little as she comes to the door to greet me, welcomes me with open arms. She loves Englishmen, fascinated by the way we talk. She's pretty, has a small, well-proportioned body not difficult to love.

At Universal Studios dubbing theatre I put sound to picture. I take sound track and negative to the lab for printing. A married print will be ready for viewing in a few days. I go home to Jilson. There's never anything to eat in her house; pills, dope, plenty of booze but nothing to eat. She's concerned about her figure, has to stay slim to get into her skimpy groupie clothes. We don't eat; we get stoned and stay stoned.

In bed watching the news on television, a shoot-out in a house down in the flats not far from here, a bunch of revolutionary kids being exterminated by the LAPD. They fire fusillade after fusillade into the house until it catches fire. Only in America.

Today I awoke to cacophonous noise. There's a sound system in every room, each of them playing a different rock album. Somewhere in the background, a washing machine rattles away. Jilson, fully dressed, moves briskly around the house tidying up. It's four in the afternoon, she wants me out of bed; she's expecting a visitor.

I stagger to the shower feeling like shit. A visitor? By five o'clock I'm in Schwab's down the bottom of the canyon on Sunset eating breakfast thinking it might bring me round. Who the hell is Jilson's visitor? She asked me to not come home until eight. Is there a lover I don't know about? Not that it matters. It's her business, not mine.

I drive back up the canyon and park down the road from the house, there's a Porsche in the driveway. I smoke while I wait. A man leaves the house, smart suit, ponytail, earring, briefcase. I wait for him and his Porsche to leave and park in his place.

Jilson is in her robe, drunk. She's out of it. It takes her a moment or two to recognise me, drools a greeting. She says she loves me.

"Come over here honey, talk English to me."

I notice a hundred dollar bill on the table. She picks it up, hiding it crumpled in her fist. I'm put out, don't ask me why. She can see I'm put out.

"A gal's gotta live, don't she?"

There's nothing I can say; we're ships that pass in the night. Near to tears, she unfolds the bill and tears it into small pieces; looks to me for forgiveness.

"Shit Mal. It don't mean diddlysquat."

I stayed on with Jilson until the lab released the final print of the movie.

107.

I'm back in Santa Fé for the wedding with a day to spare. We watch the movie and gain a degree of satisfaction from it. It's not the greatest film ever made but has its moments. Even Candice likes it.

I try to find Mike and Barbara. I ask Barry. Last he'd heard, Barbara was pregnant. They'd gone back up to New York, to Queens, to Barbara's parents. No one has an address or phone number.

Bob arrives from London and the party gets started. All of us had been married and divorced one time or another, sour experiences that fail to prevent the flow of stoned, sentimental optimism and bonhomie oozing out all over us. It's a wedding, darn it! We're going to have a good time and Barry and Candice will live happily ever after, ain't that the truth?

Surprisingly Barry chooses Brad the killer instead of me to be his best man, Bob will give Candice away. Displaced by a murderer, I'm relegated to wedding photographer. It's a civil ceremony. Barry, Candice, Bob and Brad, stand before the judge, the same one who'd bailed Brad and who will later try him for murder. They swear to honour and obey. I take snaps outside the courthouse, a few close-ups of the girls. Not Brad's wife. I daren't go near her.

The reception took place at a hotel out in the desert, a dude ranch. We take over the restaurant, about twenty of us. By the time we arrive we've all smoked a lot of grass, done a few lines of coke, no one's feeling any pain. Barry's as composed and impassive as a soldier, Candice stumbles a couple of times on the way in. Bob and I sit next to each other.

The meal is sumptuous but no one's hungry. We get stuck into the champagne and when that's gone, the wine, the brandy and cigars. As the tables are being cleared of untouched dishes, Candice stands, swaying, glass in hand toasting absent friends and we all join in. She names them one after another and we drink to them, one by one.

Pulling herself together, she becomes serious, respectfully proposes a toast to friends who are no longer with us. Dead friends. She names a couple and we drink respectfully to their memory. She's searching her addled brain for others that have died. Che Guevara. We drink. Tania. We drink. Janice, Jack Kennedy. Marilyn Monroe. Candice appeals to us, slopping her wine. Has she missed anyone?

Everyone's trying to recall anyone who's died that we have forgotten. Suddenly I remembered Michael Cooper a fellow photographer who'd topped himself. Candice had modelled for him once or twice and I thought she might remember him. I call out his name.

Candice freezes. She turns slowly in my direction, lowers her arms dropping her wine glass. The room falls silent.

"Michael? Michael Cooper?"

Long pause.

"Michael's dead?

I nod, dumbly. She fixes me with venomous glare. She's reacting like poor Michael was her bosom pal. She hardly knew him.

"You chose my fucking wedding day to tell me Michael Cooper's dead?"

Volcanic anger accumulated over millennia is erupting and headed in my direction. She wants to get at me, tear out my throat. How can you anticipate this kind of thing? She's overturning drinks and chairs in an effort to reach me, Barry and Brad holding her back. Bob and I look at each other, shrug and go outside for a smoke.

Bob and I flew back to London together in contemplative mood. It was the wedding. Candice had taken things to a new low and Barry was playing along. I'd been surprised by him not asking me to be his best man, choosing a stranger instead simply because he was a murderer. I suppose I should be happy for him whatever his choice. It was his show. What does it matter? But it's taken something away from us. We'll never be quite the same buckaroos again.

108

Back in London in Bob's flat he suggests we go and collect the Land Rover, the one he'd been given by that poor young sod of a coke dealer who'd fallen victim to Candice. It was parked somewhere in Ibiza. He has the address.

I'd never been to Ibiza before, Mediterranean ghetto in the sun for lotus-eaters. English, American, Swedish, a few French and German, all very friendly, all stoned and ready to party at the drop of a hat. I'm not in the mood for partying and Bob's preoccupied, keener than a kid to find the Land Rover. It's where James had told him it would be. It's in mint condition, fires up right away. Bob's delighted with it. He likes its military feel, its heavy-gauge steel, its impregnability. We drive around the island until sunset, then onto the ferry bound for Barcelona. We put up in a hotel.

In the morning we head for Paris. Bob has some business to attend to; it's on our way; a guy he's done a few deals with. Bob says he's good people and has a beautiful old lady. They live in Loupian, a village outside Narbonne, a hundred kilometres over the French border. Meeting new people, being sociable at this time has little appeal for me but I don't object. I'm thinking about getting back to Wales, back to work. Not anxious, but it's on the mind.

Loupian is by the sea. We arrive at the house around 10 a.m. Bob knocks hard on the door but there's no response. He knocks harder and starts yelling. Someone's coming. The door opens and there, standing with a towel wrapped around his loins, is Tim not believing his bleary eyes. Bob had no idea Tim and I had met before. Tim's excited to see me again, both of us amazed at the coincidence. He leads us into the bedroom. Cheetah is waking up and as amazed as Tim to see us. We sit around on the bed. Tim chops up some coke and we all do a line or two.

Wide-awake from the coke, Cheetah is chattering on about their exploits with the Living Theatre. In telling a story she has this way of making you feel you really missed out if you weren't there. Bob obviously doesn't care. He's doing more coke, Tim's back in bed smoking a joint. Cheetah hasn't stopped talking, first about the Theatre's politics, now she's on about their libertarian sexuality. Bob is unzipping his pants and climbing into bed between the two of them. I've seen enough. It's not yet eleven o'clock in the morning. I go for a walk on the beach.

Tim is surprised when Cheetah tells him she's going to Paris, taking a lift with us. I hadn't heard about it either. She must have arranged it with Bob. Tim's upset. What's to say? I've been there myself and know how it feels. Bob is as compassionate with him as you can get but I can see he's quietly delighted by the turn of events. He really fancies Cheetah. I feel sad for Tim as we wave him goodbye, so forlorn suddenly without his mate.

Cheetah and Bob chatter away on the drive north. I'm charmed by a French accent. I'm hearing her sexy, nasal pronunciations but thinking about Wales and writing again. Thoughts are returning to the story where I'd left off. Summer's gone and it's time to work. It was late evening when we checked into a hotel on the outskirts of Paris.

Bob books the rooms, adjoining doubles. We dump our bags with the concierge and find a place to eat. After dinner we return to the hotel and go to our rooms. The door between them is open. Bob has moved Cheetah's stuff into his room presumably intent on continuing what he'd started with her in her bed that morning.

I'm lying on the bed in the other room chatting with her. She doesn't really want to be back in Paris, she's just looking for a change. She asks me what it's like in Wales. Bob is out of the shower. I can see him though the open door, drying down in front of the mirror. I tell her how it is, deep in the heart of the countryside, so distant from main highways few visitors ever make it down there. Bob is deodorising his armpits. It's an isolated life, not everyone's cup of tea. Bob pokes his head in, grins at Cheetah. She gets off the bed, walks through to him and says she wants to sleep with me. Bob grunts; shrugs. She returns to my room, closing the door.

We leave the following morning. I've invited Cheetah to Wales. The mood on the trip is subdued. Bob lights up a spliff and passes it around. By the time we get to London it's business as usual.

I'd left the car parked in the mews outside Bob's flat. Cheetah and I get in and wave him goodbye. I watch him in the rear-view mirror as we pull away. He's happy to have a Land Rover. He's admiring it, looking at it from every angle.

We drive at speed along the M4. Every time I drive this road I think of *The Arousing*, zipping down here in the MGB with Billy the American actress masturbating under her blanket and dear old Mike in the back, recording. We cross the Severn Bridge, taking a road that runs alongside the Wye, past the ruins of Tintern Abbey mysterious in the dying light. Cheetah sighs at the beauty of it all.

The lights are on in the cottage, a Mini parked outside. Alex has a visitor? I draw in next to it. Cheetah sits staring at the beauty of the scene before her, pays homage to the gods that brought her here. She's like that; can't let anything pass without making you aware of how awesome it is. It's already becoming predictable. She sees a flower, dew on a spider's web, a cloud formation so exquisitely beautiful she insists if you don't pay attention to it you'll be missing something extraordinary. She's offering you an experience so precious that you're in some way beholden to her for it. She probably means well but it's already beginning to annoy. She opens the door of the car and gets out. Alex comes out of the house to greet us. She and Cheetah are excited to see each other. They haven't met since that evening in *Confusion Cottage* in Santa Fé. We go inside. There's no visitor; the Mini belongs to Alex, a gift from the godparents. While I was away she took driving lessons and has a licence.

After supper we sit around the fire catching up on gossip. Alex's heard from Mike. Barbara gave birth to a son. They're still living in Queens with Barbara's parents although they're not there at the moment. They won a lottery. They'd gone to hear some jazz one evening at the Village Gate, a benefit concert for a radical activist on the run. There was a lottery and they'd won it, three weeks in Jamaica all found and that's where they are now, they left the baby with the grandparents.

Alex says she's leaving, moving up to London, moving into a flat in Primrose Hill with another girl, someone with money. They're starting a business together; oriental antiques. Alex's going back to Nepal to buy a few things, Tankas, robes, amber beads; she has a refined taste in such things; a shop in Kensington is interested in anything she brings them.

Alex left today. Packed all her belongings methodically into the Mini, every square inch of the car loaded to capacity. We embrace and kiss. She's behaving like a loving daughter leaving home not sure I'll be all right without her while I'm slipping seamlessly from one relationship into another. The Mini bumps away up the lane, suspension almost to the ground, exhaust box clattering on every rock.

There's a lot I like about Cheetah but find her spiritual chatter irritating. She talks a lot about religion, the Bible, the Upanishads, the Seven Fold Way, always seeking paths but never finding them. When she meets someone new, she's immediately onto them about the meaning of life.

Tim calls her. They talk. He's coming over. I'm relieved he's going to resume charge of her. He arrives. We hug each other. I've prepared food but he has cocaine and a bottle of Margaux. He rolls an enormous spliff.

He's already rented a place in a village near Monmouth, ten minutes from the Mill. For a young guy Tim's well organised, everything covered, nothing left to chance. He takes an interest in my writing, suggests we might work on something together. I tell him about the aborted Oedipus western. He likes it. We'll work on it. And we did. When we'd finished I handed it to Linda who promptly sold an option to Hollywood producer Zev Braun.

109.

Tim's from a wealthy Beverly Hills family, his father a successful movie producer. Lucky man you'd think. But it wasn't so. He says it was during a serious earthquake in LA that he fully realised how unimportant he was in the life of his parents. He was about four years old suffering from a heavy dose of measles. He was alone in the house, hands tied to the bed frame to stop him scratching his spots, when the quake hit. The whole house shook and collapsed. He and the bed ended up below in the sitting room covered in dust and bits of debris; he waited several hours before being noticed. He knew, from that time on, he was on his own.

We enjoy the auspiciousness of our meetings. Without looking, we found each other in Casablanca, Santa Fé and an obscure village in the south of France. We acknowledge a sense of brotherhood. We know we will be seeing much of each other.

My brothers are all married and have children. They tell me living alone is unnatural. Man was meant to marry and multiply. I find being on my own liberating. I live in a world where everything is the way I want it. I can do what I like when I like. Selfish I suppose but perfect for a writer. No one else's needs to consider, no interferences, few distractions and the loudest sound in the world the babbling of the stream outside.

During the week I rarely see another human being. I go to the George after Saturday morning shopping and hobnob with the local gentry for an hour or two but the remainder of the week I'm working. That's all I do. The only breaks I take are for the management of the house, the shopping, feeding myself and, when the weather's fine, a walk in the woods. Sometimes I sit on a tree stump the other side of the bridge and meditate. Sitting quiet like that, all manner of small critters appear and play in front of me.

110.

I've stopped thinking about the heroin story I call *Confessions of A high Official*. I've put it aside. I'm working on a screenplay of the love story I call *The Secret Room* and getting nowhere with it. It has structure; a beginning, middle and end. What's holding me up is the main character, the poet, the way his mind might respond to salacious suggestions from his students' letters. I have to reveal his sexuality, his desires, his fetishes. In a synopsis you can get away with it, get the idea across in a few generalised words without having to commit to detail. In a screenplay every action requires description in explicit detail the way it'll be seen on the screen.

In creating a fictional character, my only genuine source of reference for the character's feelings, thoughts and behaviour is myself. Revealing my own libido is not easy for me. I don't really want to go there. Being sexually honest and writing it down for all to see is proving to be a problem. I don't say anything to Linda but I'm not sure I can deal with it. I'm disturbed enough to put it aside and return again to the Confessions.

I'm in London. I drive up here about once a month to score hashish. It's available closer to home but I choose not to make connections with dealers on my own doorstep, No one knows who I am in Notting Hill.

When I'm here I try to see a friend or two, maybe spend a few hours with Bob. On this occasion I'm eating in Soho with Terry Donovan in his choice French restaurant because it claims. blazoned on the outside of the restaurant. "Le Patron mange ici"

Terry an I have been friends since I first commissioned him to take photographs for me when I was an Art Director in advertising. Now he's a director producer already with one movie under his belt.

As usual the story I'm writing is ever on the mind. No matter where I am or what I'm doing, the story's being churned over. All it takes is for Terry to ask what I'm up to and out it comes. I tell him about the *Confessions*. He listens in silence as I tell him the story as far as I've taken it. He shakes his head the way people do when they're amazed how certain they are of something and says it's going to be his next movie. He takes out chequebook and pen, writes out a cheque for a substantial amount and passes it to me; "a wave of the flag". He interrupts with raised hand my protestation that I can't take his money as I haven't yet figured out the ending. "Say no more". He's that certain. He'll talk to Linda.

111.

I'm back at the Mill; a post card from Jamaica. Barbara writes she and Mike have never been so happy. It's paradise and they never want to leave.

A phone call from Alex. She's going back to Kathmandu for a month, a shopping trip. She's throwing a leaving party and insists I come.

I'd love to see her and wish her luck but I've a lot on my mind. Agent Linda has been asking about *The Secret Room*, how is it coming along, wants to know if she can read some pages. There are no pages. I've been avoiding it, I can't write it. I panic when I think about it and I'm not in the mood for a party. I tell her I'm sorry; I'm at a critical phase in the writing and can't let go. She sulks, says it's a rotten shame I can't take a few hours off to say goodbye to a dear friend. After all these years, what we've meant to each other, what we've been through together. She could lay it on as well as anyone.

She shares a flat with Charlotte, her business partner. It's a large semi-basement garden flat in a Victorian square off Regent's Park Road, white painted, pine-floored, a few oriental prayer mats spread around to slip and slide on. It's sparsely furnished with little on the walls, a few Tibetan Buddhist pieces. By the time I arrive there's already a large gathering, young, bright, voluble, balancing drinks and plates of food.

It's good to see her again. She's looking more beautiful than I remember. I'm glad I came. I've known Alex ten years, a long time in a young person's life. It makes me feel old to have known someone that long. I've watched her grow, chrysalis into butterfly. She hasn't metamorphosed into a whore like she did in *The Arousing*; she's become a fast lane butterfly.

112.

I'm not much for standing around talking. I've little to say these days and nothing that makes for polite party conversation. When I find myself in a crowd I've developed the practice of disappearing. The first step to becoming invisible is being silent. I find a comfortable armchair at the back of the room where I'm unlikely to be disturbed. I sit, like I do on the tree stump outside the Mill, gazing at the young critters playing their social games.

From behind the chair, hands are stroking my hair. I let it go on for a minute or so. Then a low, slightly accented female voice says my name. I turn slowly, fantasies chasing each other around the mind a billion a second until I'm facing her. Roughly cut, short, dark hair, intense eyes, no make-up, denim shirt and jeans; none of it belongs to the voice; none of it meets the crazy mind's sexual criteria. There's no spark, no flash of lightening, no chemical explosion. She smiles confidently.

"I'm Su."

She's in her late twenties a student director at the National Film School. We chat. I ask her what she's working on. She talks quietly, rapidly with a slight antipodean accent. Says she's finishing up a movie, her graduation film. She asks what I'm working on.

Although I'm engrossed in the *Confessions* it occurs to me she'd be more interested in *The Secret Room*. I need to talk to someone about it; she's a director and an overtly sexual person. I tell her it's a love story. She comes out from behind the chair, settles herself at my feet.

It takes but a few minutes to give her the outline. She's fascinated by it. Sexual fantasy is an area she's been exploring in her films. Would I like to come back to her place? She'd like to discuss it with me. It seemed vitally important I talk to her. She's obviously an intelligent, purposeful woman and I'm curious to know what she might bring to the screenplay but despite her sexy accent I'm not interested in spending the night with her. I make my excuses.

She's undeterred.

"What about tomorrow?"

We agree to meet the following morning around eleven in the French Pub in Soho.

I'm standing at the bar waiting for her. I glance occasionally at the mirror behind the bar watching the door, looking up every time I hear it swing open. It's a quarter to twelve. I've been waiting three quarters of an hour and about ready to go.

The door swings open. I look up; an attractive woman comes in. She looks around. I'm on my third large whisky, not a lot for me but I'm warming up. The woman takes her place next to me at the bar close enough to allow the suspicion of a delicate shade of perfume to pass between us. I can't be expected to wait all day for this Su person. I cast a sideways glance at the newcomer. She's looking at me from beneath mascara eyelashes, cigarette between immaculately rouged lips. She smiles, takes the cigarette from her mouth, fingernails lacquered, and says in low, accented voice,

"You going to buy me a drink or what?"

113.

Su? Is it Su? I don't believe it. An extraordinary transformation! The short, untidy hair permanent-waved into soft curls, the lids above her intense eyes shadowed, mascara on lashes, short, close fitting, velvety dress, seamed nylon stockings, high heels. She can see I'm impressed.. She can see she hit my mark. I buy her a drink, we laugh and chat and one thing leads to another.

We're in her flat, a dismal, sparsely furnished room in Shepherds Bush. Busy, untidy desk, typewriter, telephone, answer machine in between overflowing ashtrays, soiled plates and glasses from long forgotten meals. Stove with pots and pans, empty bottles on the floor, beer, vodka; humming refrigerator, clothing slung over furniture, over the open bathroom door.

I'm driving back to Wales. How did she know? Who would consider such a thing? Changing persona to seduce a complete stranger patently not attracted to her. I don't know what to make of her.

She behaves as if she's an actress auditioning for a role, seeing herself as Nadja, the femme fatale in my story. Perhaps she can bring the much-needed sexual fantasy to the screenplay. I'm enjoying a sense of adventure again, a tingling apprehension at the prospect of this arcane story of my own imagining spilling over into real life with Su. Her student project is in post-production; she'll be free in a week.

I admire women and enjoy their company. They bring out a gentleness in me that gets discarded when I'm with men. I fuss around a woman, help her on with her coat, carry her bag, bring her flowers, I don't think about it, it's the way I am. When Su arrives at the Mill, I go to help her off with her coat and she shrugs me away with a look of impatience; she can manage. Of course she can manage. She doesn't need me to help her in and out of her coat. Why do I continue to make these absurd gestures of *politesse*? I seem unable to override these learned rituals of gentility. The way she responds suggests I'm being insincere. But I'm not. I'm paying court to her, to the mystery.

The I Ching was written five thousand years ago, a three dimensional commentary on every aspect of human emotion. It's cleverly structured to advise you, at any given moment in time, what's in store for you down the road. It cross-references your present situation with likely outcomes and what you need to consider if you want to avoid negative results. Carl Jung took a great interest in it. It sat comfortably with his own notion of synchronicity, the sense that all events are consciously connected. Su wants us to consult the oracle together to see what five thousand years of Chinese wisdom has to say about our budding relationship.

114.

To consult it, we cast three coins six times, symbolically freezing the moment in which they were cast. The way the coins fall, heads or tails, guides us to a particular passage in the book. The chapter to which we're directed tells us we're undertaking a perilous journey. It warns that water extinguishes fire and fire evaporates water. Fire and water? All astrological identities are represented by an Element, mine by Fire, hers by Water. We look at each other with complete equanimity, undeterred. After all, people are known to survive perilous journeys.

We haven't started to discuss the screenplay. She's content to relax and absorb her new surroundings. We go for long walks, further than I'd go on my own, sometimes five or six miles. I introduce her to the town, Saturday morning shopping and the local gentry at the George.

I know, from having seen her apartment, there's nothing house-proud about her. I'm particular when it comes to the kitchen, food left lying around, tops left off jars. I mention it to her. She says if I want the tops back on jars, I should put them back on.

She's a hopeless cook, strangely disinterested in the process, doesn't have the patience. The tea leaves the pot before it reaches the cup. Food to her is simply fuel for the engine prepared in as short a time as possible. It doesn't matter what it is or how it tastes as long as it's reasonably wholesome. I do the cooking for what it's worth, with no more enthusiasm than her.

115

We chat about filmmaking, suspension of disbelief, what it's all about, why we do it, does it have a purpose. I tell her I relate to film as an artist relates to a canvas. I tell her the way in which *The Arousing* was conceived. She'd like to see it. I have it on VHS and show it to her after dinner.

I haven't seen it myself for a few years. I still enjoy it; still think it's as good as we thought back then. Su watches it in silence to the end.

She thinks it's an *Onanism*, a wank. The poetry, the abstract values are lost on her. She's brought a couple of films she'd like to show me, films made last year at film school.

116.

The first one is short, not even ten minutes, two talking heads, a young man and an attractive young woman. They're playing scrabble. He asks if she's married. She says no. He's encouraged. She doesn't want his attentions, she just wants to play scrabble. After a few more moves she adds she used to be married. He shows interest. Divorced? No. She's a widow. How sad. He's sympathetic; he'd like to hold her hand. She says her husband committed suicide. The man doesn't know how to deal with this information, slowly withdraws his hand and returns his attention to the scrabble.

She enjoys his discomfort. She continues, tells him she'd been at pains to figure out where she was, what she was doing the moment of her husband's death. She'd been in the US travelling across the country on a crowded Greyhound bus. She'd gotten chatty with the man in the seat next to her. After a few hundred miles and a bottle of vodka they got to kissing and cuddling. It was night there, early morning here. She'd worked out that while her husband was dying in London, she was in this bus somewhere between New York and Chicago and a complete stranger was coming in her mouth.

The young man freezes, the scrabble piece in his hand poised over the board. The cool juxtaposition of these events is too much for him. He can't look at her. He puts down the scrabble piece, stands and walks out. End of story.

The second film is a more ambitious production, an erotic piece about thirty minutes long. It centres on a young working-class lad dissatisfied with his job as a Post Office clerk. He wants a more glamorous life. He has dreams and ambitions he thinks could be realised by becoming a stand up comedian like they have in all the workingmen's clubs around the country. He tries out his act in front of his mum, his sister and his girlfriend. He bombs. His mum and sister think he should stick to the Post-Office.

His girlfriend doesn't think his act is that bad. She thinks it's a question of presentation, suggests it might be improved if he were to do it in drag. She says it's quite the thing in the workingmen's clubs these days, drag artistes. She thinks he'd make a beautiful woman.

Buoyed up by her enthusiasm, he agrees to do it. She brings him some clothes. He gets into a garter belt, stockings and knickers. Putting on a slip, he finds the brush of the silky fabric against his sex arouses him to a spontaneous orgasm. He can't help himself.

He gets to like wearing women's clothes, begins to prefer them to his usual gear. He makes his preference plain to his girlfriend. She can't deal with it. What she started has gone too far for her. She dumps him.

Mum is mystified, confused, her working class mentality unable to deal with a transvestite son. She wishes he'd never had the idea. It's lucky his father's not alive to see him dressed like a girl. Only his sister is sympathetic. She's helping him dress when he has another spontaneous orgasm and she holds him to her like a lover. She encourages him to be who he wants to be, steals underwear from Marks & Sparks for him, takes him out dressed as a woman, walking arm in arm with him in public. When flirty boys whistle at them, he senses the power he has as a women that he didn't previously have over these men. He doesn't have a problem with his metamorphosis, he's happy the way things have turned out. Upbeat ending.

The production of the films is amateurish. Students work on nil budgets. But the content, the sexual honesty, had the impact of truth. It tells it the way it is, whether you like to admit it or not. In *The Arousing* I'd wrapped my own libidinous meanderings in poetry, everything meant but nothing overtly stated. That's what was good about it. Even so it's meaning was blatant enough for the Lord Chamberlain to ban it from public exhibition.

Su's films are quite different. They engage without shame with real and radical relationships. I know what she meant when she said her work explored the kind of territory we were about to enter.

117.

Su first talks about the screenplay on one of our long walks. There are numbers of areas that need developing. We must start by creating in the mind of the poet Garcia Aureliano something that might later give rise to sexual anxiety and mistrust of women, a childhood memory perhaps, a traumatic event possibly involving a sexually dominant mother. Su wants to concentrate on the poet's fantasies. We'll act them out; see what plays and what doesn't. I'm to be the poet Garcia Aureliano; she'll be Nadja the dark lady of his imagination. As she's to be Garcia's creation she'll behave exactly the way he wishes, do to him as his lover the things he fears and desires. Without a trace of a smile she says, "OK? You're Garcia. I'm Nadja". "What do you want me to do to you?

Confrontation. This is it. I'd never yet dared enter those forbidden rooms of erotic imagination. I'd hoped she would lead me there and now she's at the door inviting me in. The confidence is being tested; it still mistrusts the situation. We walk on in silence, Su impatient for an answer. Half a mile further on, in all probability recognising my shyness she says,

"Bondage is fun. D'you think he could get off on bondage?"

"Bondage? Sure. Why not?"

Garcia could get off on bondage. What the hell; it's only a screenplay."

Su's face lights up.

"The students could suggest it to him in their correspondence; send him erotic pictures from porn magazines of people in bondage. Let's see how it plays."

She turns the sitting room at the Mill into a film set, closes the curtains leaving the room dimly lit by candles and the flickering fire. We undress. I'm naked on the divan warm by the fire. Su, with seriousness of purpose, binds my wrists, arms, knees and ankles. I complain she's being over-zealous. Ignoring me she pulls the bonds tighter. I shout. She gags me.

She stands naked, contemplating what she'll do next. I watch and begin to sense the excitement. As if enacting a ritual she brings a mirror to the light, props it on a chair, kneels in front of it impassively making up her mask-like face. She hangs glittering pendants from her ears, a necklace. In the shadowy candlelight, she sits astride me, rocks back and forward until my muffled protests of rape culminate in a violent and mutual orgasm. I'm exhausted. she unties me. I'm laughing and trembling from the experience. She holds me close, whispers in my ear.

"Shall we write it in? "

118

Su also wants to explore one of the themes she played with in her graduation film; the female latent in the male psyche. We've hinted at Garcia's neurotic relationship with his mother. Perhaps, deep in his psyche, he would like to be her, look like her, dress like her. We'll try a scene.

Su, in the role of Nadja, tells me,

"Garcia, there's a dear friend I'd like you to meet; a girl friend. I've invited her to visit. She should be here quite soon."

Garcia grumbles.

"I don't want visitors; I want to spend my time with you alone."

She ignores him.

Draping all the mirrors in the bathroom she gets me into the bath. Bathes me, shaves me. Out of the bath she makes up my face, eyes and lips; brushes out my long hair into a more feminine style; dresses me in lingerie; slip, stockings, a satin robe all borrowed from a friend of hers in the BBC's costume department. The silky fabrics are pleasantly strange to my skin. She undrapes a mirror for me to see her work, the friend she would like me to meet. I see this sexy female that looks a lot like me, looking at me from the mirror. Su, aroused by her own creation, is fascinated. She can't keep her hands off me, caressing, kissing.

119.

The screenplay's looking good; scenes are beginning to come together. Su's infused it with a sexual honesty I was too shy to bring to it. It's changed our relationship. All the petty life-style aggravations that bugged me about her are all forgotten, swept away in a tide of sexual excitement. Linda calls relieved to hear the work is moving along. I tell her a first draft should be ready in a month.

We're now regulars at the George on Saturday mornings, on first name terms with all the local gentry. They take an immediate interest in Su and my company is in greater demand. On my own they see me as a threat to their hegemony but I'm welcome when I bring a woman with me; she's fair game. We get invited to dinner parties. When, some times, too drunk to drive and we stay overnight, we find people trying to climb into bed with us. Perhaps we're exuding a sexual energy generated by the writing. Perhaps it's Su's particular magnetism. She enjoys sexual power over men.

120.

Su's gone. The screenplay's finished and delivered and she's gone. She has other commitments, a friend in New York, an old lover. No promises were made. She's gone. It wasn't that I didn't expect her to leave when we'd finished. She'd made it clear from the outset that there were other people in her life, other projects. Everything feels suddenly empty around the Mill. I have the *Confessions* to complete but can't focus on it, can't release my mind from the past few weeks, can't stop thinking about Su.

Before leaving for New York, She attended her official graduation at the film school. The invited speaker was Karel Reisz, an acclaimed director. She took the opportunity to hand him a copy of *The Secret Room*. He promised to read it. A few days later he called me (the only telephone number on the script). He was outraged. He'd never read such rubbish. People don't behave like that! Were we some kind of perverts? He went on and on about it. I couldn't get a word in. Finally he slammed down his phone. In a way his reaction made it clear the script hit the mark. It disturbed him and he wasn't up to it. Too bad.

Immediately after that I get a call out of the blue from Mike. I hadn't heard from him in a couple of years. He sounds like I feel, his voice heavy with the meaningless of it all.

What's up?

Barbara's dead.

I don't want to hear this. She never did get to leave Jamaica; OD'd in her island paradise.

He's in Queens with Barbara's folks and the kid and not feeling so good himself. Listening to him in his sadness, remembering how we met, all the craziness we shared over the years, I'm thinking we're all dying and not going to see each other again.

I mope around the house nursing my depression. Linda calls. She loves the script, says it's fantastic, the best thing she's read in years. She has a client waiting to read it.

I have the script of *Confessions* to write for Terry. He wants me to reconnoitre the places where the movie will be shot. Good historical and topographical research will reinforce the authenticity of the story. An envelope arrives by special delivery, two thousand pounds cash and a round-the-world first class ticket. I'm back on track. I have places to go and people to see. I'm operative again, heading first for Washington DC. Most of what I'm looking for is there on film at the National Archive.

121

I'm in a hotel around the corner from the White House. I've registered at the National Archive and have already spent a few days reviewing America's twentieth century history recorded on hundreds of thousands of feet of film. New deal Roosevelt, both World Wars, Pearl Harbor, the Cold War, Vietnam. It's the greatest movie show on earth; American footage, British footage, German, Russian, Japanese footage. I'm fascinated. As I watch I'm structuring out of it a plausible stage for my fictional characters to make their entrances and exits.

One evening I return to the hotel from the Archive and there's a message waiting for me from Su, a telephone number in New York City. I go to my room and call her. It's good to hear her voice again. She got my number from Linda. She'd like to see me. I tell her to come down to Washington and join me on my trip. We can work on the *Confessions* together. She says she'll be there in the morning.

We're going to Hawaii. You'd think, the way we're behaving, we were going there on our honeymoon. I have to keep my mind on the work. We're going there so I can stand amidst the memorabilia of ill-fated Pearl Harbor and soak up its history, the surprise attack by Admiral Yamamoto's dive bombers and torpedo planes one sunny Sunday morning. We have a stopover in Los Angeles. Su has friends in Hollywood. She'd like to visit with them. We're in no hurry.

122

Sherri's a movie producer. She and Su met in London when she was an Assistant Producer on a movie being made at Pinewood. They'd met in town at a gathering of feminists, quickly became friends, then lovers. Sherri's obviously delighted to see her again, her dark eyes fairly sparkling with recollection. They're soon off in the garden together whispering.

Sherri's doing her best to ignore me, behaves as if she's alone with Su as though I'm not there at all. She's in the middle of a production, her first feature. It's only a 'B' but she insists Su watches tomorrow's shoot. She'd invite me along but it's a sex-sensitive scene, has to keep crew and bystanders to a minimum.

The following evening, she and Su are getting ready to go and see the rushes. Sherri's very apologetic; it's a very small viewing theatre. I look at Su. She shrugs, puts her arm around Sherri's shoulder and they leave. On my own with nothing to do, I flick through my LA address book and there's Jilson. I telephone her. She's ecstatic, surprised, speech slurred; can't wait to see me, hear me talk English. I leave a note for Su telling where she can find me.

Hollywood for me is at its most beautiful in the cool of the dying day and full of memories of the times I'd spent here. Things were different then, Al Vandenberg, Michael the kid, Alex and Jilson; we were Gypsies parked on the outskirts of the known universe. Now I'm comfortable in the back of a taxi, a well-paid writer in the legitimate movie industry, a different feeling.

The taxi drives east along Sunset, passed the Chateau Marmont where Mike and I first met Tim Leary with Alan Douglas. We hook a left up Laurel Canyon. I don't know how I managed to get here. What did I do to turn things around? I'm enjoying it but it's far from front line, hands on filmmaking. Things change.

123.

But Jilson hadn't changed, the same wispy spirit floating between heaven and earth on a cloud of drugs and alcohol. She stumbles to the door to meet me, shoves her pretty, smiling face up to mine, a big kiss, drags me inside and pushes me into an armchair, sits on my lap, lights a joint, takes a deep drag and hands it to me.

She babbles on and on with alcoholic difficulty about the good times we've had and how much she loves me. Wouldn't I like a drink? Stumbles across the room to a table laden with booze, pills, dope of all kinds, pours two full glasses from a half gallon flagon of vodka, slops them as she passes one to me.

Fear and loathing in Laurel Canyon, both of us drunk and stoned in a number of directions. I can hear Jilson still mumbling in my ear how much she loves me. I'm the only one she ever really loved. Talk English to me Mal and then Su's there.

Perhaps I'm hallucinating. No, she's there all right. I try to introduce her to Jilson but my tongue won't work. Jilson, comfortable on the floor next to me, becomes aware of her presence, looks up at her, squinting against the light. Words barely audible, she wants to know who is this stranger? What does she want? She offers her a drink.

Su picks up the flagon of vodka and empties it over Jilson and me. Takes me roughly under the armpits, forces me to stand and move to the door. She doesn't say a thing, not a word. I remember stumbling down steps and being stuffed into the back of a car.

I'm in a shower, a freezing cold shower; I've got all my clothes on. I get out. I'm conscious but not feeling at all good. I strip off soaking clothes and dry off with a towel. The bathroom is familiar. I'm back at Sherri's house.

I have no idea what time it is. I pass by the sitting room. The door's open, a dim light. Su and Sherri lying naked on the couch. I pause for a moment. Su stares blankly at me as I stagger by on my way to bed.

124.

We arrive in Hawaii, the island of Oahu. We settle ourselves in a glitzy white concrete and glass hotel on Waikiki Beach and cool out for a day. I'm still hung over. On the second day, feeling stronger, we take a cab to Pearl Harbor.

It's become a museum, a reminder of Yamamoto's attack. I'd seen film in the National Archive, Japanese footage from the air, US Army footage from down here where I'm standing in the place where it actually happened. I'm matching the topography, the wharves, docks, imagining the scene, the noise and smoke, the confusion, the wounded and dying.

We cross over the dock where the great battleship USS Arizona was berthed at the time of the raid. She's still there, spooky through twenty feet of water. We walk around the museum amongst the memorabilia. Su looking closely at photos of the destroyed fleet observes it was a very old fleet even in 1941; most of it had seen action in the First World War. In any event it wouldn't have stood much of a chance against the new battle fleets of the day. She thinks the Japanese were doing the Americans a favour getting rid of their antiquated junk for them. I discover that America, at the time, did have a fine modern fleet under the command of

Admiral Halsey, battleships, aircraft carriers, the whole shooting match. At the time of the raid they were safely, conveniently out of the way on manoeuvres. Collusion? The thought occurred.

125.

Our 747 takes off and heads out towards Japan. We are on our way to Tokyo to meet with Tamiya Jiro, a matinee idol of Japanese cinema. He's made a hundred and three feature movies, all in Japan. Everywhere we go fans mob him waving autograph books. He's to play Terumasa Mouri in the movie.

Jiro's English is good. He spends a lot of time in the US, mostly in Las Vegas hanging out at the gaming tables with a gang of notorious Hollywood stars. He'd like to be as international and notorious as they are, that's why he's crazy to have the role. He'll be on screen, beginning to end, in front of international audiences. He'll become an acclaimed star. He's very excited about it; gives me the name for the role he will play. *Terumasa Mouri*, which he says means Shining Justice. He wants to help get the project going any way he can; he confides he's deeply in debt to his patrons, he calls them his bankers but they are Yakuza. They have a proprietary interest in Jiro and want to meet me. He takes Su and I to their offices, high up in Tokyo's Ginza.

A serious-looking, middle-aged bunch of suits, they make gestures of polite welcome to Su, a lot of purring, but you can see they don't really want her there. They're not as easy about women as we are in the West. Jiro translates for me. They're polite enough, they say they like the story and wish it success. They ask a few questions about Terry, his connections, his financial backing. That's all they're interested in. The way

Jiro defers to them and their tone of voice when they speak to him makes it clear he's their man, they own him. This movie is obviously very important to Jiro's relationship with them, he really needs a winner.

We discuss the movie endlessly, the character he will play. It's the first time I've related to a Japanese in this way. I'm gaining significant insights into our cultural differences and similarities. We agree Terumasa Mouri is a man who embraces traditional values, Bushido, the way of the warrior. At the same time he's a Buddhist, a pragmatist, a very practical man with a strong sense of justice.

126.

The night before we leave Tokyo. Su has the curse and is lying up in our room with a bottle of vodka watching Japanese television. I'm sitting downstairs at the crowded bar drinking Santori whisky next to a well-dressed man of about my age. Drunk, he makes no attempt to disguise his contempt for me. He doesn't know me so it's presumably because I'm a westerner. I drink my whisky and order another. I try to communicate with him, hint at a conversation. He gives me a sideways look like I'm shit. I try a smile, he turns away muttering. After a few minutes, feeling bolder, I tap him on the shoulder. He turns, suspicious. I pat my chest, point to myself; shaking my head I say,

"This isn't me."

He looks at me as if I'm crazy. I hold one of my eyes wide open with forefinger and thumb and point into it.

"This is me."

He peers drunkenly at me, looks me right in the eye for the first time and cracks up laughing. He shakes my hand. He speaks hardly any

English but whatever he's saying sounds like a good joke shared between friends. He takes me off to a restaurant, we get drunker ending the night in some little dive trying to dance together, falling about to the plink plonk of the Mamasan's koto. She's playing and singing *My Bonny Lies Over The Ocean* for her drunken English guest.

That night of childishness taught me. Beneath every culture lies the same raw material. Japanese or English, we're all subject to the same human emotions, we love and hate, fear and hope, laugh and cry. We're all life's gamblers one way or another.

127.

Su and I are on the Bullet Train to Hiroshima eating sushi from neat bamboo packages bought from food vendors trucking up and down the carriages. Like Pearl Harbor, Hiroshima has its museum. It's on the bombed site's epicentre, right where it happened, where Mankind first dropped an atom bomb on itself.

I don't think Su and I spoke a word the whole time we were there. The exhibits, the shadows of human beings stencilled onto chunks of concrete wall did the talking. Forensic photographs of radiation wounds on the flesh of men, women and children did the talking. An eyewitness is recorded as saying the explosion was like the light of ten thousand suns and the sky turned brown.

Both Hiroshima and Nagasaki, the site of the second bomb, were modern industrial cities. That's why they were targeted, no doubt about it. Their destruction was a severe setback to Japanese industry and commerce after the surrender, after things had returned to some kind of normality. They really were the losers.

We're going to Kyoto. Jiro has arranged for us to stay at a traditional Ryokan run by three women, hard futons on the floor, sliding doors of translucent rice paper. We relax into the comparative quietude of this ancient old capital of Edo Japan. We wind down.

We make love again. We haven't made love since leaving Washington. Our friendship weathered the few days at Sherri's, survived the incident at Jilson's place. There's been no criticism, no accusations. Our relationship is firm. The fantasies that originally brought us together, that set fire to our lives, are giving way to a more fundamental reality.

We wander the city; the Buddha is everywhere. We visit him in his many temples, sit for a few moments here and there in his cool shade. We contemplate the simplicity of Zen, listen to Mozart in a coffee shop. We walk for miles climbing the forested mountains outside the city. At night we make love.

Restored and ready to move on, we return to Tokyo and say goodbye to Jiro. He takes us to the airport in a limousine. Fond farewells; certain we'll meet again. We board a plane to Thailand. We're about to explore my character William Harper's domain, where he lived and the territories that constituted his sphere of influence. Neither of us talks much on the flight.

128.

Bangkok. This is how the story goes. A hundred years ago, William Harper the First sent his mule trains here from his valley in Burma, a two or three week trek in those days, mule trains loaded with raw opium fresh from his fields in the north. His partner Lord Dunsmuir, by then his father-in-law, would be here waiting with a fast clipper to ship it to Europe then on to the United States.

His influence grew with his success, first in the hills amongst Chieftains and Warlords and by the time the opium got to Bangkok he had persuasion all the way to the sea, sometimes in official places when there were obstacles to overcome, sometimes in the palaces of Princes and Kings. His last business venture, his swan song, was in partnership with a prominent German pharmaceutical company, Bayer, to produce a derivative of opium, a panacea, a miracle pill to be sold over the counter at the local chemist shop for anything that ails you, a wonder drug named Heroin.

There's no opium in Bangkok any more, only heroin, one hundred per cent pure Chinese white heroin. It's cheap, easy to find, the bellhop in the hotel will get it for you. The other choice euphoric of the region, equally easy to buy, is Buddha Weed, a strong and pleasant, locally grown marijuana, just the tender buds of the plant neatly woven round a split stick of bamboo.

The hotel we stay in is full of young Americans, demobilised detritus of the Vietnam war, repatriation postponed on account of not yet being ready to face up to the old soap opera they used to call home, preferring to stay here in easy reach of calming drugs. They're a pleasant enough bunch, always around the swimming pool, sunbathing, laughing, joking, sipping beer, smoking joints. Some read studiously, some do Yoga, some meditate. One or two I spoke to went, when the war was over, into Buddhist monasteries, spent a few years in retreat from the nightmare they'd just been through.

In our room on the second floor, the blinds are drawn against the bright afternoon sun. Su's feeling queasy. I'm smoking heroin. There's a knock at the door. We ignore it. Another, more urgent knocking. No one moves. The knocking continues. Su sighs, drags herself off the bed, opens the door. It's Alex for chrissake! She flies into our room and declares we must be crazy staying at this hotel. Didn't we know?

Didn't we know what? We haven't seen her for nearly a year, thought she was in London, Kathmandu or somewhere and here she is charging into our room in Bangkok. How did she know we were here?

This is an infamous druggy hotel. Gets raided regularly by the police. We've got to get out, immediately. I've never seen her so dramatic, so paranoid. I try to calm her down but she won't have it. If the cops find just a smidgen of dope on us we could be looking at ten years. Get out! Now!

It's difficult at the best of times to figure truth from paranoia, and quite impossible when stoned. I'm not at all sure of the next move and look to Su. She's already stuffing her few clothes in her holdall. I do the same. We follow Alex, moving at a fair clip.

I'm curious to know how she found us. She has a business appointment with an Englishman who was supposed to have arrived two days ago. He'd stupidly booked himself in here. She's been coming by every hour or so to see if he'd arrived, finally made the clerk show her the register and saw our names. She takes us to a hotel along the street above a bar where she's staying; says it's safe from police investigation. How does she know these things?

She's been here a month and knows her way around. She came to Thailand looking for antiques but all the local dealers are hip to the value placed on their treasures by Westerners. Their prices are too high for her. She's full of hope, waiting to meet this man coming from London who wants to put money into the business. She spoke to him two days ago on the phone, he was leaving right away, gave her his flight number, hotel, everything, but the fucker hasn't shown. She's wasted a lot of time and is practically out of money. She's not happy. Almost in tears she asks what we think she should do.

Su's not good at comforting people, not good with sympathy. She can see Alex's reverting to helpless dame, victim of fortune, but she's not buying into it. She tells her to use her initiative, find her own way out. Alex's tears roll. What's she done to deserve this? Thought we were her friends. I put an arm around her. She nestles her head against my chest, sobbing. Her tears, the smell of her hair, take me back to times almost forgotten. Su looks at us, shakes her head. She leaves without a word, slamming the door behind her.

I comfort Alex, father and daughter again. Her fear subsides, we chat about old times and even manage a laugh. I tell her Su and I are going up-country for a few days. If the business connection she's waiting for hasn't arrived by the time we return we'll sort something out. I give her some money.

129.

Su and I are on a bus going north to Chang Mai, the traditional market place for opium. Since our character William Harper introduced the crop to northern Burma in the 1870s, hill tribesmen for hundreds of miles around have brought their harvest here to sell, kilo upon kilo of sticky, black opium. We spend the day there, visit a temple or two then head north again to Chang Rai, a smaller town nearer the Burmese border.

Before leaving England we'd tried to get regular visitor visas into Burma but had been refused for no ostensible reason. Everyone's got their problems. The only way into this forbidden territory is by river and then a long walk. We needed a guide.

The moment we step off the bus in Chang Mai a good looking, polite young man latches on to us, says he'd be happy to show us around the temples and so on. He's half Thai, half Burmese, about twenty-four, speaks good English, name of Thomas. He looks OK to me. Su nods. We agree a fee and ask him to take us to a café before we start the tour. The three of us sit drinking tea; Thomas asks where we'd like to go first, what would we like to see? Su says,

"Poppies. We'd like to see poppies."

Thomas looks at her, then me, then her again.

"Poppies?"

"Yes. Poppies. Opium poppies."

Thomas cottons on. Big smile.

"We don't grow poppies here. You have to go into Burma."

Su smiles. I explain we're doing research for a movie and need to suck up some local atmosphere. Will he lead the way?

He's silent for a moment, frowns, thinks. Says it's all a bit illegal, even dangerous. Danger not only from the police and border guards, there are bandits around, unpleasant fellows that would kill you for the clothes on your back. I show him some money and he finally agrees to take us into Burma. He has to run home to get a few things for the journey; he'll be right back. He returns a few minutes later carrying a small rucksack. We stand to go. He holds his jacket open, gives us a flash of the automatic pistol tucked in the top of his trousers, and then we're off.

There are no roads north of here. Everyone gets from village to village by river bus, long, narrow boats powered by outboard motors plying the length of the river. Thomas leads us down to the jetty; we stand in line waiting for a boat to take us upstream. When it arrives, we board and sit in line between the narrow gunnels. The journey is pleasant. It's a warm, sunny day; the boat rides smoothly in the flat water, everyone silent, just the putter of the motor. We stop at each village, some passengers disembark, others get in.

Arriving at the border, we get off the boat and start walking. We climb out of the valley, into the endless rolling hills of the Shan province of northern Burma. In 1884, William Harper settled about two hundred miles north of here having found the soil and climatic conditions he'd been searching for and planted the seed he'd brought from India. It was a revolutionary move, introducing a cash crop into the area, introducing the notion of money and profit to the local tribes folk. It started as a local affair but by the time he died, most of Shan province was under cultivation, a successful economy tucked away up here in this remote corner of the world, between Tibet, China, Laos and Thailand, the Golden Triangle.

When his son William Harper II inherited it, he encouraged its growth, extended cultivation into China's Yunnan province increasing production. Unlike his father he sold his opium to anyone who would pay the price, to the heroin factories in Hong Kong and as far afield as Shanghai. Moral rectitude was not an issue for him. Good and Evil were two sides of the same coin.

Su's tugging at my sleeve. Thomas walks at quite a pace and she's finding it difficult to keep up, she's tiring, needs a rest. We pause and sit. It's unlike Su to give up. The sun is setting. Thomas says we're not far from the village where we'll spend the night. We get to our feet and plod on.

We arrive at the village, a cluster of houses raised above the ground on stilts. Thomas finds the Headman and arranges for us to stay the night in his house. Our elderly host shows us a place we can lay our sleeping bags. We enjoy a light meal with him and his family then crash, we're all exhausted. Our hosts retire to another corner. Soon I can smell the sweet aroma of opium smoke. I whisper to Thomas I'd like a pipe. He gets up and goes over to the family, murmurs, returns with a pipe and some opium. He prepares the pipe for me; I light up and take a deep drag. I offer the pipe to Su but she's already asleep. I offer it to Thomas. He defers, says opium's for old men.

Morning. Su is outside, throwing up. After a cup of tea she says she's OK and ready to go. I'm concerned. She's been off colour since we arrived in Bangkok. She says she's fine. She's pregnant.

One minute I'm writing a movie about power and corruption, the next I'm thrust into an entirely different world of parenting and children. If it's not one thing it's another. I've been carefree the last few years, the idea of children never crossed my mind, never seemed to be part of the plot. Su says it was our last night in Kyoto, that's when it happened, she's sure of it. The love between us that night must have been too much to resist for a wandering soul seeking a way back into the world. She treats her pregnancy as if it's a logistic, another aspect of the production to be taken into account. I tell Thomas I'm going to be a father. He's delighted, takes it as a good omen. The news revitalises us, reenergizes our backs and legs. We still have the day ahead of us and walk at a fair clip.

We're crossing fire-blackened fields. Thomas had warned us we were too late to see the poppies in bloom. When the pods of the poppies have been bled of their precious sticky sap, the villagers torch the dry stalks, the ash of the old fertilising the new.

Clouds cover the sun, there's a nip in the air. I've seen enough and Su's tired. We make it back to the village where we spent the night on our way in. They greet us like old friends. We eat with them. When we bed down, our host, the old Headman, comes over with a pipe, chuckles, hands it to me and lights it.

130

Back in Bangkok Alex has a boy friend, a young American and they're having a good time. He has plenty of cash, takes her out to expensive restaurants, buys her expensive clothes. She pauses long enough to welcome us back, kisses us profusely when we tell her Su's pregnant but saddened when we tell her we're on our way back to England. She says she wished she was coming with us. Why does she want to leave?

She's leaving but not yet, still has a few things to do, business things. She brightens up, whispers confidentially even though the three of us are quite alone in our hotel room. She's going to carry some Buddha weed into the US, twenty kilos. She's doing it for the boyfriend; he's giving her five thousand dollars. Easy money. Wearing the new clothes he's bought her, she'll walk through San Francisco customs on New Year's Eve like a queen. No one would ever suspect. Who's going to take any notice of a pretty, well dressed girl on New Year's Eve?

I'd heard about this kind of drug pimping, men hitting on single girls, showing them a good time, making love to them to seal the deal. The girl is now a mule, that's what you call someone who smuggles dope for you. Alex is a mule. What can I say? She's not a kid any longer. She's old enough to place her own bets, take her chances like anyone else. Five thousand dollars against two or three years in jail. Su has nothing to say either. She's surprised Alex has the guts to do it.

I'm planning to do a little smuggling myself through the mail, a few sticks of Buddha weed, a gram of skag. The following day Alex comes with us to the airport to see us off. She kisses Su on the cheeks then hugs me tight. I wish her luck.

Su's silent most of the flight home, not morose, not at all. She's calm and silent. When we arrive at Heathrow she says she's going to her own flat, wants to be alone for a while. I still haven't got used to her independence; I'm not comfortable with it. The taxi drops her off at her flat in Shepherds Bush and takes me on to Paddington Station. I'm going back to the Mill. I board a train to Bristol Parkway where I'd left the car. It's sitting in the car park; been there these three months. It clatters and coughs as I start it then bursts into song. There'll be no food in the house so I drive into Chepstow on the way and shop for the essentials. The sun's setting as I bump down the track to the Mill. I'm knackered.

There's a pile of mail sitting under the porch outside the kitchen door. I scoop it up, take it into the house with everything else and dump it on the kitchen table. There's a parcel from Thailand addressed to The Misses Penelope and Dora Faversham. I rip it open. Two cylindrical cans of tea I'd mailed in Bangkok. I take off the tops. In each, under a layer of tea, fifty sticks of Buddha weed. I rummage in the bottom of one of the cans and come up with a small glassine envelope, a gram of pure Chinese white heroin.

131.

The house is freezing. I switch on all the heaters, build a fire, make myself a meal, open a bottle of wine, put on music. After I've eaten, I roll up some of the Thai grass, sit in front of the blazing fire smoking, listening to some memorable rock and roll that drifts me away from an anxious present back to earlier times.

I woke unusually late this morning. It's been a long summer of sun and endless blue skies. Now, through the bedroom window, rain-heavy clouds move fast across the grey sky skimming the tops of the pine trees. I watch a bird struggle against the wind. I whisper to it to let go, let the wind take you. The 'phone's ringing. It's Linda checking to see if I made it back. She's on top of the situation at all times. There have been 'phone calls from Terry Donovan. He's looking forward to the first pages.

132.

I call Su but there's no reply. I try again later in the day. Still no reply. I call Terry. He's not there. I leave a message and try Su again. Nothing. I drive into Monmouth to augment supplies, enough for the next week or two, and then back down to the Mill. I check the answer machine. No messages. The weather is closing in. Barely four in the afternoon and it's almost dark. I light a log fire in the sitting room and sit in front of it sipping whisky, not at all sure of my next move. The responsibility for writing *The Confessions* weighs heavily on me. The mind is on other things.

God alone knows what's happened to Su. Always an unpredictable entity, now a pregnant unpredictable entity, there's no telling what she might get up to, where she might go. You know how the mind runs on, churning over the darkest of possibilities.

The following morning I try 'phoning her again. This time I get a disconnect signal. I think of driving up to London to her flat but decide against it. She probably forgot to pay the telephone bill and they've cut her off. If I arrive and she's there I'll have to face her usual scornful rebuttal of my concern. She'll get in touch when she's ready. I call Bob. His phone is disconnected. Where the hell's he gone?

Terry calls. I tell him the trip was successful. I have all the background covered. I'm already working on the script but finding it difficult. Have I got a title for it?

"*Confessions of a high official?*"

"Hmmm. Yeah; a bit long. Considered *The Golden Triangle?* More commercial you know, easier to remember. He's tried it out on a few people, got a good reaction."

I tell him I'll give it some thought. He goes on about it but my mind is elsewhere. Where the hell is Su? Why doesn't she call? Terry senses my distraction. Asks if I'm OK. I say of course I'm OK but there's a problem. *The Confessions*. It's a huge canvas and I'd welcome help from a writer with experience of epic form. Terry's not pleased to hear this. I shouldn't have agreed to write it if there was any doubt I could finish it. He's not happy. He'll talk to Linda. Hangs up.

Linda calls. Terry's been on to her. What did I say to him? He sounded worried. Is everything OK? No, Linda. Everything's not OK. I can't handle *The Confessions* on my own. Why not take it as far as you can then we can discuss it with Terry. Yes, Linda. I'm working on it right now, as we speak, as soon as you get off the fucking line darling, as soon as the bitch Su calls.

You sure everything's OK?

Of course, sweetheart. Everything's just fine.

I put down the 'phone.

133.

I go upstairs to the bedroom and take out my drug paraphernalia. I sort out the Chinese white heroin, give myself a shot and relax. I remain in this limbo for a couple of days shooting a little dope, waiting to hear from Su, unable to put my mind to the script. Now in the mornings I'm beginning to feel the need for a shot, the body vaguely uncomfortable until I administer to it. A shot sorts me out physically but worries me at the same time. I look at the amount of heroin still remaining. It looks untouched. A gram of pure is about an ounce when cut to usable quality. I was looking at an ounce of street heroin. Even though I like to think of myself a master over addictions, when this lot is finished I'll be out on the street looking for more. I tip it down the lavatory.

I don't feel any better. I still can't concentrate on the script, still aggravated by Su's non-appearance. I feel negative, unproductive, waiting for something to happen, waiting for something to push reality along into a more satisfactory, more comfortable phase.

I lay in bed late into the morning, my mind wandering the decade of the life just passed. What was it all about? Where is everybody? Are Barry and Candice still together? Are they alive? I don't know. What's Bob up to? Kelly is dead, poor young James the coke dealer is dead, Barbara is dead, Mike didn't sound too good last time we spoke and Alex's about to walk through San Francisco customs with twenty kilos of Buddha weed. The phone rings. It's Su. Will I pick her up? She's at her flat in Shepherds Bush.

I've never seen her so calm, so at one with herself. None of the impatience, nothing irritates her. Her belly is growing and she's enjoying a state of bovine grace. She smiles at the countryside as we drive back to the Mill. She offers no explanation for her absence or lack of communication and I don't ask. It's some time before she talks at all. And then, as if we're picking up on an interrupted conversation, she asks how the work is progressing. Had I considered the ending? How does Terumasa Mouri finally let go of Harper? I don't tell her of my conversations with Terry and Linda.

With Su back at the Mill everything slides into some kind of order, there's a sense of purpose again. My own confusion recedes; muddled thoughts begin to refocus as I get down in an attempt to finish *The Confessions*. We're going to have a child. In seven months or so there will be three of us. OK. Bring it on. Bring it all on. Let's get on with it.

134.

I call Barry's parents in London to find out if they know where their boy is. Barry answers the phone. He's happy to hear from me; he sounds cheerful but I detect weariness in his voice. I invite him down to the Mill. He thinks it's a good idea. We'll swap lies over a bottle of tequila. We make a date.

He's drunk when he arrives in a sports car driven by an attractive thirty year-old woman, Elizabeth. His hair is short, his clothes simple. No earrings, no jewellery at all. Elizabeth is chatty and amiable. Su, still riding her state of grace, greets her like a sister.

Su hasn't met Barry before; she'd only heard my stories. She looks at him for a few moments as if recollecting what I'd told her about him. She smiles before taking his hand. Barry is charming, kisses her cheek, congratulates her on her pregnancy, flourishes a bottle of Tequila. It's soon open but Su's not drinking. She's tired from the excitement and retires to bed leaving us with the Tequila, staring into the fire.

Barry tries to sound optimistic but there's defeat in his voice. He seems to be creeping back into the private shell of his former self, the way he was before Candice. He tries to sound enthusiastic, talks about getting something going, a film maybe, but he's unconvincing.

We're half way through the bottle and Elizabeth has fallen asleep. Barry rambles on. At times he'd thought he could ride Candice, rock and roll with her chaotic emotional energy. In the end she was too tough for him. One night in their bedroom she produced her 9 mm Walther and fired at him. The bullet went well wide of him into her closet through every garment she owned. He couldn't deal with it.

Their divorce still isn't settled but she's already making wedding plans with a very wealthy older guy, head of a broadcasting corporation in Chicago. That's all she ever wanted, it was always her game plan to capture a solid and wealthy man but she was too easily distracted from her true purpose by men in her life that showed recklessness. She finds it difficult to resist a man prepared to enter her insanity, someone who'd go all the way with her. The only problem she had with the reckless ones, they rarely accumulated wealth and Barry was no different. I think Barry endured the relationship for the reverse reason; he thought Candice would be the one to lead him to wealth. It played as important a role in his fantasy as it did in hers. They couldn't see it but it was always a doomed situation. She recognised it before he did and gave him a hard time on his way out.

He feels empty. His emotions need a rest. He finishes the bottle, says he has another in the car. I suggest he stays a few days but he isn't up for it, has to get back, has things to do in town, people to see. He gently wakes Elizabeth. We'll stay in touch. Elizabeth, barely awake, is friendly but too tired to talk. She puts an arm around Barry, her head on his shoulder, eyes closed. She obviously adores him.

I'm glad they visited. Seeing him again even in his present state of drunkenness has cleared the air for me, blown away a few cobwebs. I still haven't asked him why he chose the murderer rather than me as his best man.

135

Rainbow Rosie was born at midnight on the 23rd July in Hereford Hospital; at midnight on the cusp of Su's astrological Cancer and my Leo. She's a very self-contained, well-contented baby. I have never loved another person the way I loved her. I was proud of her, wanted to show her off. I sang to her, danced with her, wrote poetry to her. Even before she could talk she would answer my doting with a smile like the Buddha. She quickly became the most important part of my life. I carried her in my arms wherever I went.

I tried to refocus on *The Confessions* but couldn't. It was a greater task than I had imagined and Terry had not been giving me any help with it. He'd paid his money, now he expected a script. He was sympathetic; he understood the distraction of a first-born but felt that I should be more professional and put the work first. He was right of course. Su is pregnant again. There are decisions to be made. She wants to have the baby in the bosom of her family in New Zealand. I give up the writing.

Her father was an air force navigator during World War 2. He flew many missions with the Royal Air Force as navigator in Lancaster bombers. He fell in love with one of the Women's Auxiliary Air Force, a chauffeur to the Wing Commander. They married and a year after the war ended Su was born.

Her mum and dad decided to put war-torn Europe behind them and find somewhere quieter to bring up a family. New Zealand. That's where Su and her sister Katie grew up, went to school, had their first love affairs and entered adulthood. That's where Su wants her child to be born. There's no need to talk to the family about it yet. It's at least three months before the baby is due. I've told Linda about the problem I have with the writing. The screenplay is far too big a project to handle on my own. Real life has taken over and making demands. She's not pleased.

I call Barry in London hoping to see him on our way to the airport. His mother answers the phone. She's bewildered. Didn't I know? They left last month for Nairobi. Barry has joined a film company there that makes commercials. He's doing very well. The money is very good and his most recent film is up for an award. His Mum was certainly relieved. She'd been worried sick over Barry, wondering how he was going to sort out his life.

If it's not one thing it's another. Alex calls. She's in jail in California. Didn't make it through customs at San Francisco airport. The judge must have taken a liking to her and only gave her eighteen months when she was expecting three years. She'll be out in nine months with good behaviour. It's a soft touch and she's learning how to service the internal combustion engine.

I'm amazed at her composure, her realistic view of the way things have turned out for her. Perhaps it takes traumas in her life to teach her what it's all about. But when I tell her the plan to go to New Zealand to have the baby she starts crying over the phone suddenly feeling the isolation from loved ones. I don't tell her about Barry.

We arrive in Auckland, New Zealand. Su's seventy-seven year old father is there to meet us. There's an air of modesty and conservatism about him, his clothes, his hair, his clean-shaven face, unsmiling but friendly enough. There's an old fashioned air about him, a certain awkwardness around a female. He leans forward to kiss Su's cheek without even glancing at her enormous belly.

A widower, he takes us to his home in a northern suburb of the city. He's retiring, set in his ways. He tells us, without much show of regret, he's unable to care for us himself, it's too much for him and that Katie will arrive soon to take us to her home in Napier on Hawke's Bay to the south of the Auckland Island. Su shrugs. It's OK. The journey towards the birth has begun, the child in Su's womb our navigator.

Katie is married to Les Spinks and they have three children. They live in a substantial house they built themselves on the outskirts of Napier. Les is a contractor. He's pleasant enough, has a good business, votes Conservative, is patriarchal and occasionally cheats on Katie. Even though he's gone from the house all hours of the day, she's too busy with children, schools, cooking, cleaning to have the time to play around. We're made to feel at home. The kids love us. We're fresh from England, a land they've only heard about.

One evening after dinner, after the children have gone to bed, Les asks Su what are her plans. Where is she going to have the baby? She deliberates. She looks from Les to Katie then back to Les, smiling.

"I'd like to have the baby here Les, amongst my family."

Les pauses in the middle of breaking off a piece from a bar of chocolate. Looks across at Katie, then at me. He looks back at Su, says,

"I don't think that's a very good idea."

Su looks at her sister. Her sister is biting her lip and shrugging, helpless.

Les adds,

"There's an excellent nursing home just down the hill."

Being born in a local nursing home was not the way Su wanted her child to enter the world. We have to move on.

We land at LA Airport. We'd called ahead and Sherri's there to meet us; she shows a certain diffidence towards her extremely pregnant friend and lover. She's full of excuses. We're not going to be staying with her. A friend of hers, a really nice guy, is lending us his house in Topanga Canyon. He would be happy to help out. We can stay there. The following day Sherri takes us up there to meet him. We follow where this unborn child leads.

Armand Levy is a lawyer from a wealthy Hollywood legal family in the movie business, a pleasant and intelligent fellow. He welcomes us warmly and makes tea. His house is in the Rich Hippy style of architecture made of wood, smooth plaster and tiles, full of light and originality; the bathroom is a rock pool with pebbles, moss and grasses and plants climbing the walls.

Su wanders around the house, her eyes alight with inspiration. She smiles at Armand and tells him she would like to have her baby in his beautiful house. Armand is charmed and says he would be honoured. Sherri says she's going to film the birth. Armand asks when the child is expected. Su's not absolutely sure, three or four days? A week, maybe? He says he has to go to Washington for a couple of weeks and he'll likely not be here. Do we have a midwife? We hadn't thought about it. He's perfectly content for us to have the baby here but insists we organise a Midwife immediately. There are several in Topanga. He has to cover himself. He's a lawyer.

We consult with a young woman, Sharon Greene, recently returned from voluntary service in the depths of Venezuela where she was midwife for an area covering many villages. She's birthed scores of other people's children but hasn't had one herself. Su's satisfied with her credentials and not knowing when exactly to expect the child she suggests Sharon moves in with us until the birth, there's room for her in the house.

Armand is a divorcee, his refrigerator devoid of food. We borrow his car and drive to Hollywood to shop for victuals at the Health Food Store on Melrose. As I park I see, stapled to a telegraph pole outside the store, a flier with a picture of my friend from early New York days Shri Swami Satchidananda. He's in town. I have to see him. I call his ashram and ask to speak to him. He comes to the telephone delighted to hear from me and insists we come over right away as he's leaving in the morning.

His ashram is in a quiet neighbourhood of large Hollywood houses off Sunset going east. When he's not there, which is most of the time, it's run in Vedic tradition by six young devotees, three boys, three girls, under the authority of an equally young but presumably more experienced novice Swami. All of them are in their early twenties, all celibate. They hold their Swami in very high regard and as his friends we are accorded a most respectful welcome.

We drink tea and gossip with him. He says he'd seen Bob quite recently. Bob had stopped by the ashram in Virginia on his way west to Santa Rosa to help a friend build a house. I try to figure out whose house that might be, our individual horizons had broadened to the extent it was difficult to keep sight of each other. Swami says he was in good health and light of heart, moving on after the loss of his wife. He's sure there will be someone else in his life. I don't doubt it.

Time for us to leave. He kisses Rainbow, kisses Su, lays hands on her belly, blessing her and the child she's carrying. He gives me a hug and I hug him.

136.

Because Su is unable to come up with better than approximate due dates, two of which have already come and gone, Sharon insists on examining her to check the clinical state of the pregnancy. Rubber gloves and much groping.

Su, sweet as can be says, with understanding, there's absolutely no need for her to be concerned. She had been through pregnancy before; her body tells her everything is as it should be and we must all be patient. Sharon backs off but becomes nervous.

Armand has gone and won't be back for ten days. Sherri is using the house as her production office for the film she's going to make, on the phone to cameramen, editors, sound engineers.

Sharon has developed a rash and stepped up her consumption of Vodka. Su knows what she's going through, her feeling of frustration at not being permitted to exercise her professional judgment and the anxiety that goes along with it but she's unable to help her. She's not about to allow Sharon to examine her in order to satisfy her disquiet. After another two days of deepening depression, her mind doubtless full of the personal and professional affront and the possibility of dark consequences for which she will almost certainly be blamed, Sharon finally blows. She shouts at Su she's had enough. It's not her responsibility any longer! Have your baby in a fucking hospital! I'm out of here!

It takes a few days to find a suitable replacement. We still haven't got one by the time Armand returns and he's nervous. He knows all the reasons why he should be nervous if something goes wrong. We find a midwife, a middle-aged, family woman with a lot of experience but even her reassurances don't calm Armand's fears. He's worried about Su's term. He tries gently to persuade her, as he would a client, she should let someone else examine her. A doctor. Su's peaceful, idyllic birth scenario is disintegrating.

Fire brakes out in the Canyon and changes the entire pitch of our lives. It happens most years in Southern California's canyons with lesser or greater severity. This year it's a big one in Topanga. A pall of smoke from burning forest is already drifting into the canyon; neighbours are on the roofs of their houses wetting them down with hoses. Higher up the canyon dwellings have already been consumed. Within the hour the Sheriff is at the door commanding us to evacuate.

We're packing up, not yet knowing where we should move to when the ashram telephones. They've heard on the radio about the fire and invite us, if we have no other place to go, to come and stay with them until it's safe to return. Armand retires to his small beach house in Malibu, too small for the entire circus. We call Sherri and she drives over, picks us up and takes us down to the ashram.

Swami has gone. Behind the house there's a tidy, grassy yard with an Avocado tree. Against the back wall of the yard, a garage converted into living space. This, for the time being, is our new home. Not the palace we've just come from but a completely Spartan, whitewashed room, a wooden chair, a straw-filled palliasse large enough for the three of us on the concrete floor, a small framed photograph on one wall of an ascetic with a heavy cast in one eye. We'd bought food to contribute to the evening meal. We cook, eat and then sit and chat for an hour or so about the nature of the universe before retiring to these monastic quarters.

Morning; we hear on the radio that the fire in the canyon is under control and people are returning to their homes. We pack and are ready to move back when Armand calls saying under the circumstances he'd rather we stay where we are. Su isn't in the least phased. Nothing seems to disturb her mood of relaxed self-confidence. She smiles at Armand's reaction.

I remain positive in the face of the anxiety we seem to be creating in the minds of everyone we meet, first in New Zealand and now here in California. It follows us around. We drift ahead on the gentle tide of Su's needs and feelings. I follow her direction. I'm a participant in her show. She'll let me know when I'm required on stage and what I have to do.

I ward off growing consternation on the part of our six young hosts and their novice Swami as it dawns on them they are lumbered with us, an unforeseeable result of their generous invitation to save us from the fire. The novice Swami is a young man already familiar with responding to changing circumstances; he seems to have some understanding of the meaning and the meaningless of it all. But his six younger wards, less experienced, share the middle-class fears of their parents about birthing a child and how dire the consequences if not carried out in the safety of the maternity ward of a hospital. That evening after dinner they lecture me for not having made arrangements. They accuse me of gross irresponsibility. Su must go to hospital.

They responded as I expected, mimicking their parents fears that something might go wrong and the medical and legal consequences that would follow. Like Armand, not wishing to deal with possible consequences real or imagined, they want us to move on. They are sympathetic but feel it would be best for all if I booked Su into a hospital

There's no way Su is going into a hospital but I can't tell them that. Even though they're young and inexperienced it's time for plain talk. I wait for the right moment in the conversation then say quietly,

"It's God's will brought us to you. We thought the child would be born in New Zealand, then it was Los Angeles, then Topanga but we've been wrong all along. He's landed us here with you."

They quieten down for a moment. I've softened the edge of the debate but only for a few moments. They still aren't happy about it. We're involved in convoluted argument about what exactly is 'God's Will' and the nature of Fate and Karma when Su gets to her feet, puts Rainbow in my arms and walks silently away. Unable to reach agreement and everyone tired, the decision is postponed until morning. The meeting disperses; Rainbow and I join Su in our monkish whitewashed room.

137.

It's a hot night; Su's taken off her clothes and lies on the palliasse. The covers are wet around her. Thinking she'd pissed herself I ask is everything OK. She says she'd had enough of the talking. She's been to the corner pharmacy up on Sunset, bought a pint of caster oil and drank it. Her waters have broken and she's going into labour. This is it. This is what the past nine months have all been about. We're approaching the timeless moment, the miracle, the arrival of a brand new surprised and confused individual into the World to begin the cycle of life.

Su is supremely confident. Her confidence is my confidence. She gets up, leans against the wall, commands me to massage her back. I massage, she moans. She lies down again. In my mind I'm merging with her, conscious of no one else.

Over a period of two or three hours, contractions happen with increasing rapidity. Rainbow is disturbed by her mother's cries and shouts. She starts crying. I pick her up and carry her to the house. I wake up the novice Swami and hand her to him. He understands immediately.

I return to Su. She gets up, leans against the wall, I massage her back, her moans are louder. She lies down again. Wide-eyed, she whispers it's coming. At two o'clock in the morning the top of a head appears in her ever-expanding vagina. I'm positioned between her legs like a scrum half waiting for the ball. The baby, a girl, squeezing herself with ease out of Su's heaving body, drops into my hands. She's perfect. A Star is born.

Like filmmakers do, we'd made contingency preparations. A pair of new tennis shoe laces and scissors sterilised and stored. Any boy-scout manual can tell you how to tie off the umbilical cord in two places with the laces and cut between them. It doesn't do to be squeamish. One more contraction and the placenta slides out. I bury it in the back yard under the avocado tree.

Early in the morning our young hosts, no longer worried by the negative, downbeat thoughts of the previous evening, come into the room to pay their respects and wonderment at the miracle God has bestowed on their house. They wish to give the girl the name Shakti to sit along side the one we will choose. Star Shakti.

Birthing remains for me a fundamental and ordinary experience, an ordinary miracle you might say. Women are giving birth every day of the week, every minute of the day all over the place, ordinary and miraculous, the core event in the ongoing survival of the species. That Su and I were on our own at the birth gives it special meaning for me. Together we experienced something awesome and beautiful. Less than eight hours after Star Shakti entered the world, the four of us are in the health food store on Melrose getting food for lunch at the Ashram.

138

When we get back we find Tim and Cheetah waiting there to greet the new baby, the whole family. It was almost Biblical. How did they know where we were? We accept their unannounced arrival at this time as just another incident in an ongoing life full of mystery and the unexpected. So how did they find us? They were in LA on account of business. Tim had to deliver some cocaine to someone in Topanga, an old school friend, Armand Levy.

Where did this movie begin? When did it take over and add its peculiar colour to the real life? Where is it taking us? How does it end? They ask what we plan to do. Neither Su nor I had given it a thought, too consumed by the present to have considered the future. We have no plans. They ask why don't we come with them. They're living in Arizona. They have a house in Tucson. Why don't we come and stay for a while?

We'd never been to Arizona. That afternoon they whisk the four of us away with them, headed for Tucson. Our hearts and minds are too full of the birth experience to leave room for any anticipation of what we will find there.

We found care and friendship, at first staying with Tim and Cheetah and then in a suburban house loaned by a friend of theirs. This friend later found employment for me in his marijuana distribution business.

He and his partner wanted me to find a safe house in Florida for imported marijuana and thought a man of my mature age, (I was nearly forty then while the average age of the gang was less than thirty, a little too young) to conduct the business now required of me. Their set-up, conducted in military fashion, no detail omitted, fascinated me. I agreed to take the offered job and had my hair cut and my beard shaved down to a moustache.

I was flown up to Chicago, which seemed to be the hub of the business, and a large, smart car was bought for me. The car was chosen on account of the size of the trunk, big enough to pack four or five hundred pounds of weed. Later I would be asked to carry a load from Florida to St Louis, by way of initiation. When the journey and transfer of goods was successful, I was given the car.

The required safe-house had to have access for eighteen wheelers and sheds on the property to conceal the tonnage of Marijuana they expected; it had to be near the main highway north and close to the coast. When I found the right place I had to organise a try-out purchase of a few months. this deal required a Banker's cheque. It was well organised. The cheque was forthcoming and handed to the house owner. When all was settled, the family moved across from Tucson and we took up residence. The property was set in a citrus orchard and I had promised to keep it in good order while I lived there. It had a swimming pool and half a dozen screeching peacocks. After all that had been asked of me, nothing, not one bud of marijuana found its way into the empty and waiting storage space. President Nixon had paid to much attention to the security of the Georgia coast where the stuff, coming from Cuba would normally be beached.

Then came the day of my initiation when I was to deliver five hundred pounds to St Louis. I drove to a given address where I was to pick up my load. Prior to the actual event, tyres were carefully checked – there would be no spare taking up valuable room in the trunk. The rear suspension was kept level and beyond suspicion by a pump off the engine.

Before loading the trunk with the smelly marijuana, I stripped off my clothes and worked naked. Not until the trunk was full and I'd showered, did I dress again. My employers had obtained an Ohio driving license for me that had no photo or any other indication of my identity other than my assumed name William Barrington. We all had assumed names and photographs were forbidden.

Then I set off north for St Louis. A two-day drive. At the outskirts of Indiana I stopped at a telephone booth to book a hotel room ahead. While I'm in the booth I see a police car drive up and park behind my car.

Of course they wanted to know if I needed help. By this time dressed quite expensively and with my short hair and moustache I looked as respectable as an airline pilot or someone like that. I was shitting myself. The Ohio driving licence passed muster, as did my perfectly honest explanation I was booking ahead for a hotel room. Satisfied I was no felon, the police got back in their car and left. Phew. The following afternoon I found my way to an address in St Louis, welcomed by a beautiful young woman; the car was unloaded and I started back to Florida much relieved.

My employers were completely satisfied and gave me the car as a gift. Now the much more demanding job of finding a safe house was to begin.

I enjoyed the act. When an Estate Agent asked if he could call me Bill I politely corrected him and said he could call me Will. My involvement with Estate Agents had it's moments. One young fellow invited me home to shoot some pool; produced a joint and asked if I smoked. I handled it like a novice. On another occasion I was in an Agent's office when a Sheriff came in, He was introduced to me as the Agent's cousin. The sheriff was curious. I revealed I was a writer. He said he had a tale to tell. His cousin had let a house to a bunch of young people and he, the Sheriff, had visited the house to welcome the folk to his area. As he approached the house, windows and doors were flung open and the inhabitants scrambled through them and ran off into the countryside. Make what you will of that. Despite these high moments, the experience was upsetting my stomach. My bosses wanted me to stay longer, but after a year I was ready to go.

I'm in New York, having a meal with Johanna and she asks what am I working on. I tell her. She draws a leaflet out of her bag and says, 'When you're finished you should have a look at this." she brings to my attention the leaflet which was an invitation from Allen Ginsberg to join him celebrating the fiftieth anniversary of the publication of Jack Kerouac's On The Road. Of course. Everyone will be there.

As soon as I returned to LA I looked-up Lewis MacAdams. Lewis was a poet/journalist I'd met in Barry's company some time ago. I offered him partnership in the Kerouac idea. He was very keen and drew in a friend of his, Richard Lerner, who had a small production company and the equipment that goes with it, We skivvied around and got a little money from friends – the cost of taking a crew up to Boulder Colorado where the party was to happen under the auspices of the Naropa institute, a University founded by the Tibetan Chögyam Trungpa, where Allen taught in the faculty of poetry called The Jack Kerouac School Of Disembodied Poets.

Back in LA, Lewis and I checked out the interviews he had very professionally conducted and from which we wrote an excellent script adding in all the moments we could remember of Talk Shows that had praised or criticized Jack's work. We set about looking for one hundred thousand dollars to cover postproduction. Most studios and distributors when asked for the cash and participation replied, 'Jack who?"

Arlene had been my assistant to the producer and she was no slouch. She made sure she had a signed release from everyone we filmed.

We had a thick wad of them.

My friendship and partnership with Lewis came under strain. Probably the fruitless search for post-production money didn't make things any easier.

Arlene and I gave up. She had been offered jobs at the Staatsoper and the Theater Aan Der Wien in Vienna, there was accommodation for two, so we went.

Not long after we arrived in Vienna, Richard Lerner's father died – he'd been a jeweller and left Richard a sizable inheritance. He wanted to finish the Kerouac film and offered me $20,000 for the releases but he didn't want me; I think both he and Lewis selfishly thought it inappropriate for an Englishman to make a film about an essentially American writer; that's the kindest interpretation I could put on a very selfish act but the money took the sting out of the unfair play. It was my film; my baby and Jack didn't belong to anybody. He was Universal.

Arlene and I returned to London, We tried at first to live in Cornwall but it proved too far from the Dance World and too much travelling. She found a place in London and we moved in, my teenage daughters also moved in. Arlene wasn't prepared to go through teenagers again, she'd had two of her own. She decided to return to her home, New York and that was that. What a girl! She had a great sense of humour that saved us on more than one occasion. When the shit hit the fan as it did on occasions she would say, "Fuck me if I can't take a joke!"

It was not long after, I got a phone call from Aziza saying her dad, Bob, was dying and if I wanted to talk to him I had best do it now. I couldn't speak, I was completely choked. How could this be? Bob wasn't supposed to die. I tried my best to say something to him; I didn't want to say goodbye.

I have to tell Barry. I called him in Nairobi to find him in an equally poor condition. He said he was seriously unwell. It was his liver but he would not stop his steady intake of alcohol and it was killing him. He died just a few days later.

Bob had written to me a few times and had visited me in Cornwall. I knew he had a partner, an artist named Madeleine Leddy, but he always came alone. I called her to share condolences over the death of a man we had both loved. We arranged to meet when next she was in London and carry on where it had all left off. Bob died more than twenty years ago, soon followed by Barry. Dear Al Vandenberg suddenly found himself in the grip of cancer and passed from this world, a world that seemed to be crumbling around me. Most of my closest friends had gone. I never thought it would be like this.

Living back in London again I reconnected with my daughters, Rainbow, and her younger sister, Star, who had opted to live with me in Cornwall. Arlene saw them as moving in on me and, having already experienced her own two teenagers, thought enough was enough and fled home to New York.

139.

There was room to reconnect with Jean; we agreed to meet many times but most times it never happened. Then one evening it did. She came to my flat with her daughters and a Grandson. Jean told me she had been diagnosed as having Alzheimer's disease.

That was bad news enough. Then she started talking about everything we've heard about, seen movies about the kind of life one can expect living with this particular mental disorder.

Assuring me she had given it every thought, inside and out, she had decided to take advantage of an aided departure from this world. Even as I write, the tears are forming, as they did when she told me. Both her caring daughters were helping her, sharing her sadness with true stoicism. They would go with her to Switzerland and be there for her while she dies. As they were leaving and I was giving Jean a kiss, she invited me along. I wasn't ready yet. I was old and the body was beginning to play me up but I didn't suffer from anything as tedious as Alzheimer's to encourage me to take such a final measure.

I guess we all experience the sadness and frustration in due course of losing our friends and loved ones. It's so obvious. We all have a limited time to enjoy this strange unknowable phenomenon we call life; Jean didn't act surprised when it was upon her. She always was brave.

A life of unintended consequences by Malcolm Hart

Barry Hall / Kelly Wilson / Bob Mayo / photographed on Primrose Hill.

A life of unintended consequences by Malcolm Hart

Malc & Mike at Woodstock

My friendly and brave cameraman partner. Unfortunately no longer with us. I loved him and will never forget him. A man open to, and unafraid of life's challenges.

A life of unintended consequences by Malcolm Hart

Terence Donavan.

Basil

Arlene

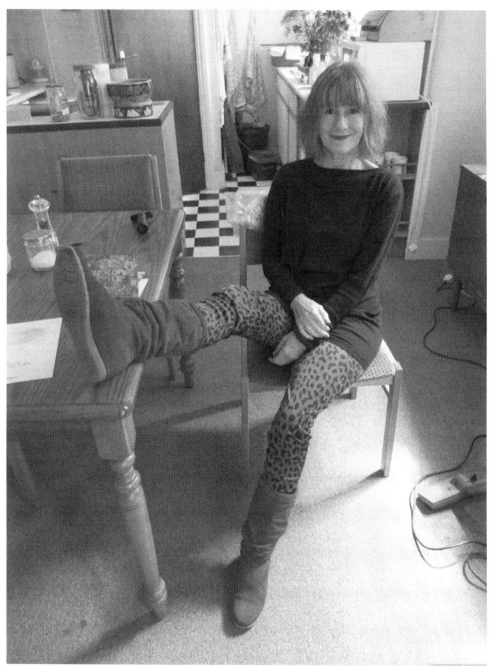

Madeleine Leddy

Printed in Great Britain
by Amazon